D1230226

BRIDGES

Engineering Masterpieces

euler_bernoulli_bea
Golden Bridge, Da Nang, Vietnam
Ice bridge,
Los Glaciares
National Park,
Argentina

BRIDGES
Engineering Masterpieces
Dan Zettwoch
IK BRUNEL
Royal Albert Bridge,
River Tamar, UK
Iya Kazurabashi Bridge,
Tokushima, Japan
30
:01
First Second
New York

Published by First Second
First Second is an imprint of Roaring Brook Press,
a division of Holtzbrinck Publishing Holdings Limited Partnership
120 Broadway, New York, NY 10271
firstsecondbooks.com
mackids.com

Library of Congress Cataloging-in-Publication Data is available.

First edition, 2022
Edited by Dave Roman and Tim Stout
Cover design and interior book design by Dan Zettwoch and Sunny Lee
Bridge consultant: Stuart Nielsen

Penciled with Prismacolor Col-Erase non-repro blue pencil. Inked with Kuretake Pocket Brush Pen and Pentel Presto! Jumbo White-out Pen. Colored with vectors in Adobe Illustrator along with my son Archie's cheap watercolor paints.

Printed in July 2022 in China by Toppan Leefung Printing Ltd., Dongguan City, Guangdong Province

ISBN 978-1-250-21690-8 (paperback)
10 9 8 7 6 5 4 3 2 1

ISBN 978-1-250-21689-2 (hardcover)
10 9 8 7 6 5 4 3 2 1

Have you ever wondered why bridges don't fall down? A big bridge weighs many thousands of tons, and yet it appears to hang effortlessly in the air across an estuary, a deep gorge, or whatever it is spanning. And then you go and whack a line of heavy trucks on it in each direction, and it still doesn't fall down—at least you hope it doesn't just as you begin to drive across it! Well, that's thanks to the clever engineers who have worked out how to make it stand up safely under all the various loads and extreme conditions it has to cope with.

But that's just the start of what engineers do. They must think about what materials to use, where those materials have to come from, and how long they will last. This is because we have to be careful not to use up the earth's limited resources, and we don't want to use materials that are harmful to the environment. And then there's the problem of how to build. Imagine trying to construct a long-span bridge out over the water—you can't exactly stand a crane out there, can you? So how can you lift the pieces into position?

Engineers use science to answer questions like, "How can I make this work?" or "What happens if I try this?" Science is cool, and without it, engineers could not do what they have to do. Using the results of scientific experiments combined with a practical understanding of physics, engineers are able to work out how to do amazing things! So, for example, a scientist might discover how to make a strong material like steel even stronger by adding extra elements or using different processes during its manufacture, and then engineers can work out practical methods for how and when to use the new, stronger material when it comes to their next project.

As someone who designs them for a living, I can tell you that every single bridge is different—because every bridge location is different! Before anyone starts building, several factors need to be considered. The engineer must decide what type of bridge is best depending on all sorts of factors, like how long it needs to be, whether it is in a city or the country, the nature of the landscape, how easy it is to build, budget concerns, and more. Also, the bridge must look good! At least, I think it should, because you wouldn't want a world full of ugly bridges, would you? I try to make my bridges beautiful so that they're not just useful but something aesthetically pleasing that people will actually appreciate, like they would a work of art! But what looks right in one place may not be the best idea in another. There's a lot to think about, and what I really love about bridge design is that it combines science with art and architecture to make things both functional and attractive.

I love the four main characters in this book. Bea, Archie, Trudy, and Spence each have a preference for a certain type of bridge, and they all get excited about how to use scientific facts and technical knowledge to come up with the best solution for each bridge. And the other thing is that they work as a team. They don't always agree (it would be very boring if they did!), but they work things out together. This is exactly how it works in reality, too. I always work in a team, and you never know where the best ideas will come from. Sometimes the best ideas can be really unexpected, and a brilliant new solution that you have never thought of before can emerge. It is a lot of fun!

So the next time you cross a bridge, you can ask yourself what materials have been used, which bits are carrying the loads, and how it all works together to make it safe for you to cross. And you can also ask yourself what you do or don't like about it or whether you think it looks good in the landscape. Has the engineer done a good job and successfully applied both the art and the science of engineering in the design? I hope you will be able to say yes to that question if you ever encounter one of my bridges!

—Ian Firth,
bridge designer

June 14, 1874, 88 feet above the mighty Mississippi River...

A daring circus owner.

A nervous steelworker.

An overconfident chief engineer.

Thousands of excited onlookers,
eager to see the crossing.

Folk wisdom says that an elephant has natural instincts and will only cross a safe and stable structure. In this case, the elephant is right!
Splendid!
Whew.
Told you so.
Railroad traffic can now cross the great river, finally connecting east and west.
OSAGE
Riverboat traffic can still safely pass underneath the bridge spans.
A new type of material is proven. The steel industry is on the rise.

Eads Bridge across the Mississippi River
St. Louis, Missouri, USA
GRAND OPENING!
SUCCESS!
Hooray!
The powerful and graceful arch design will help inspire the city's iconic landmark.

Genius!
Right? Who knew elephants were so smart? I knew about dolphins, and I once saw a chicken playing poker—
Dude. She means the bridge is genius.
Oh yeah. Look how the forces are in perfect balance!
Not my usual style, but I'd be up for checkin' it out.
Add it to the list, but we've got a lot of other bridges to cross off.
We ready?!

Today. The Ann W. Richards Congress Avenue Bridge, aka the "Bat Bridge," Austin, Texas.
I believe we are ready!
Ladies and gentlemen, our expedition begins now!
BEA
FWIP!
ARCHIE
ROLL!
TRUDY
CLICK!
SPENCE
Let's go!

Meet the
B.A.T.S.
BEAM • ARCH • TRUSS • SUSPENSION
GLOBE-CROSSING BRIDGE BRIGADE
Whoa, Archie. Dude. You been working out?
Yeah! I don't wanna be left in the dust like last time!
Oh yeah, the "Bridge of the Black Cave incident." I forgot about that.
Oh, I didn't forget.
Guys! Wait up! Huff Puff

BATS
Club motto: "WE HANG AROUND BRIDGES!"
The BATS mission: to visit and catalog the world's oldest, largest, most structurally innovative, and just plain battiest* bridges.
Pace yourself, young man!
We've got a long way to travel!
I got this! Y'all just tell me where to go!
*from "batty," /'badé/ adjective. Definition: crazy; insane. (Although, like bridges, we know bats to be brilliant and complex creatures.

Let's meet the members of our gang:
Name: Beatrix (last name unknown)
aka "BEA"
Specialty: Beam → Bridges
Personality:
Straight & to the Point
Call me "Trixie" and you're history, dude.
Engineering student
Tools:
Phone
loaded w/ calculator and drafting apps
Spotting scope
High-beam flashlight
Loves:
Lasers
Cyber-punk
Snakes
Fears: None
Name: Arcimboldo Da Ponte
"ARCHIE"
Specialty: Arch Bridges
Personality:
Jokester / Romantic
Hey, I love history!
Teen archaeologist
Tools:
Maps
Water bottle
Magic markers
Trowel & mini-shovel
Loves:
Pasta
Mytho-logy
History
Fears: Caves, tunnels, snakes

Specialty: Suspension Bridges
Personality: Enthusiastic but Nervous

BEFORE BRIDGES!

Goat Canyon, California, USA

Xiao River, Hebei Province, China

The Firth* of Forth, Scotland
*"Firth," an inlet of the sea
The Golden Gate, California, USA
Awesome, but missing something.
Like, what do you do if you need to get from here...
...to there?

You swim, take a boat, get a jetpack, ride a shark, giant trampoline–
You build a bridge! But how do ya start?
A scientist starts with an observation.
THE SCIENTIFIC
1 Make an observation and ask a question
"What kind of bridge would work here?"
Do background research:
STRONG WINDS
TALL SHIPS
DEEP WATER
EARTH-QUAKES
ETC.
2 Construct a hypothesis
Let's try a suspension bridge!
Every bridge is a "hypothesis," an educated guess at what will be successful.
Let's hope very educated.

I *observe* the other side of the bay and wanna get there. And I *don't* observe a jetpack or trampoline nearby.

Great, Bea! Let's move forward using the trusty ol' *scientific method*.

Method!
3 Test your hypothesis
Is it functional?
Is it safe?
Is it too expensive to build and maintain?
Is it beautiful?
4 Analyze data, draw conclusions, and share results
Run the experiment again if necessary.
Next time let's try an *arch* bridge.

We follow a similar method in the engineering department. Here's the:
ENGINEERING DESIGN PROCESS
1
Identify a need and research the problem
A
B
?
2
Imagine and plan a solution
FEEDBACK
3
Test and analyze the solution
4
Develop and improve your design
Finalize design and build it!
Let's remember, it's not an exact science.
Engineering = science + art.
Invention and genius are needed at every step.
BOOM!
Don't forget good luck! I heard some bridges weren't "designed" at all, just happy accidents!

SECTION
1
BEAM BRIDGES
Keep it straight.
Also!
Purposes!
Who uses bridges?
Loads!
What are they?
Forces!
Friend or foe?

Aberdare National Park, Kenya
The first beam we visit in our quest is this gigantic tree trunk that has fallen across some raging waters.
THE EARLIEST TYPE OF BRIDGE
LOG BRIDGE
Spence was right. Sometimes all you need is good luck.
The Gura, "The Fastest River in Africa"
A lucky accident like this has been used since ancient times as a crossing for people and animals.
Bridge traffic: Wildlife, nomadic pedestrians, cattle herders c. 3000 BCE
Traffic beneath: Fast fish, daring swimmers

Type: **Simple beam bridge**
Material: **Timber**
Total length: ≈ **20 m**
Longest span: ≈ **16 m**
Clearance: ≈ **8 m**
Opened: **Date unknown**
Can I get a little more luck to make it across?
Whatsa matter? Afraid of a little water?
B.A.T.S.
30
Hey! I'm not afraid! My first name means "fish" in Swahili.
I just don't wanna, y'know, get my new sneakers wet.
SPLOOSH!
Don't worry, fellas. That beam is sturdy. They don't make 'em like that anymore.
They do! High-tech bridges built today use the same basic ideas. We'll cross those bridges when we get to them.

So what is a *beam*, anyway?
As far as bridges go, it's ***the simplest structural form***.
That's why I like it. Straight and to the point.

The word "beam" comes from Old English for "living tree."
Now a beam can be made of anything, like steel or concrete.

A beam can even be made of sunlight.
FWIP
But I prefer the night.
Ooh, you're so mysterious, Bea.

You've heard of a balance beam.

My favorite is the laser beam.
But you can't walk on one of those. Yet.

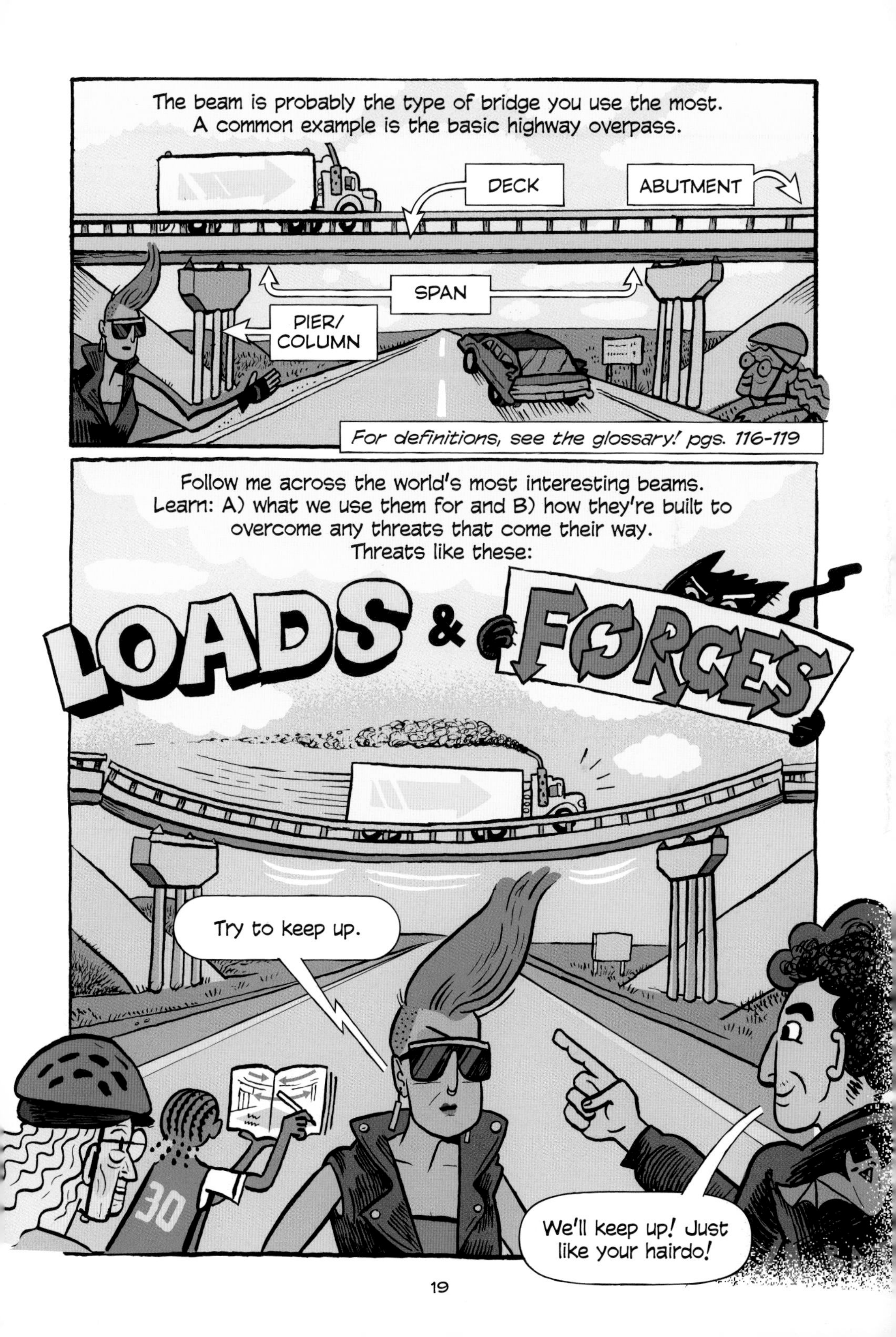
The beam is probably the type of bridge you use the most.
A common example is the basic highway overpass.
DECK
ABUTMENT
SPAN
PIER/
COLUMN
For definitions, see the glossary! pgs. 116-119
Follow me across the world's most interesting beams.
Learn: A) what we use them for and B) how they're built to
overcome any threats that come their way.
Threats like these:
LOADS & FORCES
Try to keep up.
We'll keep up! Just
like your hairdo!

Exmoor National Park, Somerset, England, UK
Next we step across some stones in the River Barle.
Ye Olde
Tarr Steps
Some say this "clapper bridge"—or "pile of stones"—is prehistoric.
Reliable sources say it's more likely medieval.
Bridge traffic: On foot and hoof
I can't believe we've already made it from Africa to Britain!
Ah, the magic of bridges, young Spence!
WINK!
Legend has it that the Devil built *this* bridge as a place to sunbathe.

Type: **Clapper bridge**
Material: **Stone slabs**
Total length: **55 m**
Longest span: ≈ **2.6 m**
Clearance: ≈ **1 m**
Opened: **Possibly Bronze Age, more likely c. 1400 CE**
This is a continuous beam with multiple spans—17 of them—supported by stone piers.
At two tons each, these dudes should be pretty sturdy.
But be careful, BATS. These stones get slippery with moss and water.
Not to mention the Devil's suntan lotion!

Quanzhou, Fujian Province, China

The BATS head far east to the mouth of the Luoyang River to cross one of the four ancient bridges of China:

Type: **Clapper beam bridge**
Material: **Megalithic stone slabs**
Total length: **1,200 m** (originally) **731 m** (now)
Longest span: ≈ **8 m** Clearance: ≈ **3 m**
Opened: **1059 CE (Song Dynasty)**
What does "megalithic" mean?
Huge rocks stacked up. Like Stonehenge.
The guy who built this bridge, Cai Xiang (蔡襄), had some cool new engineering ideas:
1 *Alternative Energy!* Workers used tide power to float the huge stones into place.
2 *Hydrodynamics!* Boat-shaped piers direct water flow and protect the bridge.
3 *Bioengineering!* Sticky oyster ooze glues the foundation to the piers, strengthening the bridge.
His ideas worked! The bridge is still here, along with that huge statue of him.

Let's take a break from the beam bridges and get heavy. Let's talk about *loads*.

A *load* is a weight or kind of pressure. There are three kinds of loads that affect bridges. The first type:

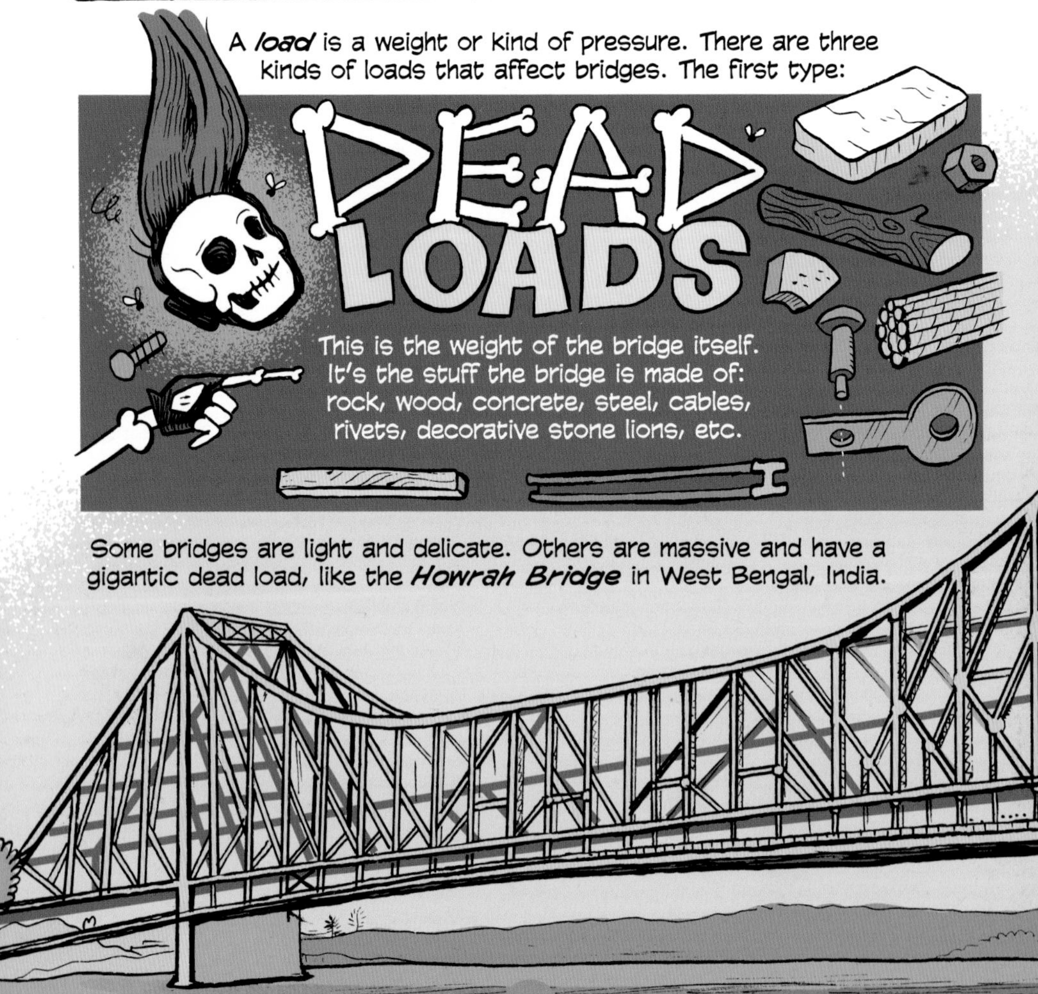

DEAD LOADS
This is the weight of the bridge itself. It's the stuff the bridge is made of: rock, wood, concrete, steel, cables, rivets, decorative stone lions, etc.

Some bridges are light and delicate. Others are massive and have a gigantic dead load, like the ***Howrah Bridge*** in West Bengal, India.

Then there are the temporary loads that move across the bridge: people, livestock, wagons full of goods, motor vehicles, etc. These are called:
GOODS CARRIER
HORN PLEASE
LIVE LOADS
Because live loads are always changing, they are trickier to plan for.
The Howrah Bridge is super busy, carrying more than 100,000 vehicles and 150,000 pedestrians a day. That's a lotta live load.
Why aren't we visiting that one?
I don't like crowds.

Bridges are often located in extreme, unpredictable places: over fast rivers, high above gorges, spanning rough seas, etc.
That brings us to the third type of loads:
BOOM!
ENVIRON-MENTAL LOADS
Uh-oh.
We're talkin' hurricanes, fast winds, high heat, ice storms, tidal waves, earthquakes, floods, the works.
One bridge that attempted to stand up to these acts of nature was the (first) *Tay Rail Bridge* in Scotland.

There are other types of environmental loads—vandalism or sabotage, for instance—but a very dangerous one for metal bridges is *corrosion*.
Among iron and steel bridges, *rust* is a constant threat.
Even the chemicals in bird poop can cause metal to disintegrate.
On December 28, 1879, the Tay Bridge was up against all three types of loads. The bridge's deck of heavy iron girders (*dead load*) was being crossed by a rumbling locomotive (*live load*) during a violent storm (*environmental load*). The rest is history.
Hey, Bea. I guess loads weaken the structure of your hairdo, too, huh?

Now the BATS are in the middle of nowhere, peering over the edge of a steep valley. They're looking at a massive wooden framework known as the:

GOAT CANYON
1126
SD&AE
1126
It was built to cross the Carrizo Gorge, part of the death-defying "Impossible Railroad."
Bridge traffic: Railroad (now abandoned)
Traffic beneath: Desert bighorn sheep
A ***trestle*** is a unique type of beam bridge supported by lots of angled legs, which distribute the weight of heavy loads.

Type: High trestle beam
Material: Redwood timber
Total length: 229 m Height: 61 m
Opened: 1933
TRESTLE
WORLD'S LARGEST ALL-WOOD TRESTLE
Let's go. What's the holdup?
Uhhh...no thanks. I'm gonna skip this one.
Really? It's just like a roller coaster.
I skip roller coasters, too.
Uh...you guys. Look at this sign.
BEWARE!
Hike the canyon and cross the trestle at your own peril!

North of New Orleans, Louisiana, USA
The BATS stay in the southern United States, but go someplace much wetter.
Lake Pontchartrain
THE LONGEST BRIDGE (over continuous water)
N
E
W
S
BridgeS, actually. They're side-by-side twin bridges.
Whoa. They are long. I can't even see land on either side of us!
This bridge is so long it's got its own radio station!
Bridge traffic: Four lanes of motor vehicles
Traffic beneath: Pleasure and fishing boats

Type: **Low trestle beam**
Material: **Prestressed concrete**
Total length: **38.4 km** Longest span: **54.9 m**
Max clearance: **15.2 m** (bascule closed)
Opened: **1956** (southbound) **1969** (northbound)
Causeway
It's got this drawbridge called a *bascule* in the middle.
Spence, thanks for letting us borrow your station wagon. It'd take us a month to walk across this bridge!
Thank my grandma. She says we just have to bring her back a souvenir.
What should we call our sweet ride? The BATSMOBILE?
That might be trademarked. How about BATSWAGON?
Too literal. Let's call it the ECHOLOCATOR. It's how we hunt our prey. (Our prey is bridges.)

Dudes. We know bridges handle *loads*. But how? Exactly what is happening when we drive a heavy station wagon across a bridge? Let's look at:
FORCES!
Like the forces of good and evil?
Kinda.

What is force?
In science, a *force* is any kind of pull or push on an object.
PULL
PUSH

A force can cause an object to move or to stop.
PFF

The object can remain in place if the forces are balanced.

The object can be *deformed* and change shape.
Uh, Archie. You might not wanna be deforming Grandma's fuzzy dice.

Force is measured in ***newtons*** (N), thanks to Sir Isaac Newton.

While studying objects, forces, and motion, he defined the laws of the natural phenomenon known as:
GRAVITY!

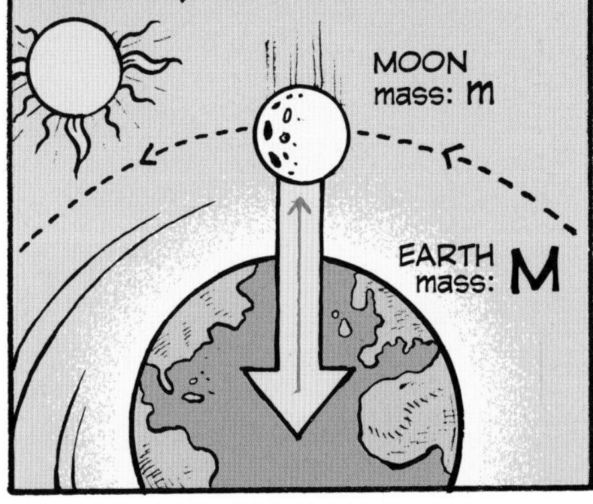
His law of universal gravitation says that bigger objects attract smaller objects toward them.
MOON mass: m
EARTH mass: M

Gravity is not technically a force. The theory of relativity redefines it to be an effect of the curvature in the fabric of space-time.
Thanks, Einstein!

Either way, gravity is always trying to pull things down.

So for bridges, the effect of gravity is the major force to be reckoned with.
Gulp

Let's take an even closer look. Besides gravity, what are the forces acting on bridges?
Analyzing forces is crucial to a structural engineer.
THUNK
Let me grab some crafting supplies.

Take this ruler. It's made of wood, just like the log bridge or the trestle. It'll be our beam.

If I press down on the ruler...
...you won't see it, but the beam is experiencing a powerful force.

The force of *compression.*

COMPRESSION

Grab the AFM.* With this dude, we can take a super-close look.

ATOM

WOOD MOLECULE

CHEMICAL BOND

*Atomic force microscope

Now when I press down on the beam you can see...

...how the wood fibers inside are being ***squished together.***

This is the force of *compression.*

When I remove the load, the force is released.

The molecules spring back.

SPROING!

We call that ***elasticity.***

I thought only my socks and underwear were elastic!
Actually, Archie, every solid is at least somewhat "springy": wood, steel, even concrete.
CARD-BOARD TUBE
Now let's make a bridge out of our ruler. Put the beam on a couple of "piers"...

When I press down, you'll see our bridge bending.
Along with *compression* on top of the beam...

...there's now another force on the bottom of the beam:
the force of *tension*.

TENSION
Remember what the regular wood molecule looked like?
Now we're looking specifically at the ruler's underside.
5
6
7
8
9
10
11
12
Now see how it appears under a bending load.
This time the wood fibers are being *stretched apart.*
This is the force of *tension.*
You can really see tension in action with a rubber band.
Or my underwear!
Dude. Gross.
When you remove the tension, it springs back.
SPROING!

Near Schwäbisch Hall, Baden-Württemberg, Germany
The BATS cruise along the Autobahn super-speed highway, and are suddenly on a high beam spanning a deep valley.
Bea, step on it! There's no speed limit!
BATS always enjoy their time on a bridge, taking time to savor its structural significance and soaking in the view of the surrounding landscape.
I'm about to puke.
I think young Spence has a touch of ***gephyrophobia***.
Uh-oh. Spence, you're on the wrong adventure.
(the fear of bridges)
Bridge traffic: Four lanes of motor vehicles
Traffic beneath: Small river, road, footpath

KOCHER VIADUCT
Type: Box girder beam
Material: Prestressed concrete
Total length: 1,128 m
Longest span: 138 m
Max clearance: 185 m
Opened: 1979
aka Kochertalbrücke
I'm not afraid of bridges. ⋟Gulp⋞ I love bridges.
It's okay, Spence! It's good to respect the forces of nature. We are soaring through the sky on a few bits of concrete.
Okay then. I'm terrified.
The bending forces in the middle of each span are the same as the ruler: the top surface gets slightly shorter (squished by compression), the bottom gets slightly longer (stretched apart in tension).
That's why the beam is shaped like an empty box: more material at the top and bottom is stronger. That's also why many beams are shaped like an "I."

Between Shanghai and Nanjing in Jiangsu Province, China
Okay. Here's the last beam bridge to check off our list.
It crosses land, water, and everything in between.
And it's got good Wi-Fi. It's...
YANGTZE RIVER
DANYANG
CHINA
N
KUNSHAN
SHANGHAI
THE WORLD'S LONGEST BRIDGE (total)
DANYANG-KUNSHAN
Bridge traffic: High-speed electric trains of the Beijing-Shanghai railway, the fastest scheduled train in the world!

Type: Beam bridge
Materials: Concrete and steel
Total length: 164.8 km
Opened: 2011
GRAND BRIDGE
丹昆特大桥
Now this is more my speed!
Even going 350 km/h, this bullet train will take more than half an hour to cross the bridge.
There are no spectacular views or death-defying parts of this bridge...
...but it sure gets you a long way fast!

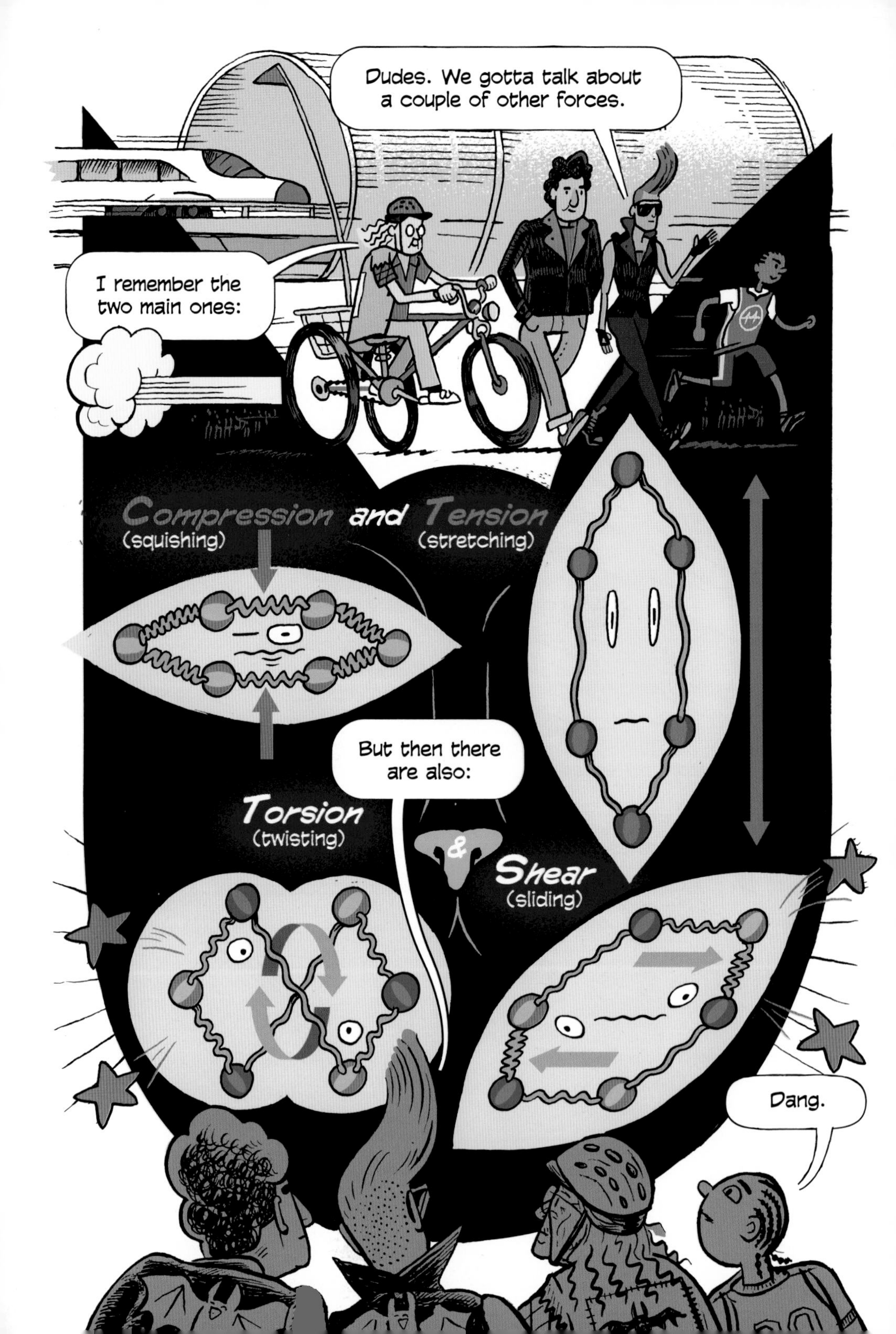
Dudes. We gotta talk about a couple of other forces.
I remember the two main ones:
Compression and Tension
(squishing)
(stretching)
But then there are also:
Torsion
(twisting)
&
Shear
(sliding)
Dang.

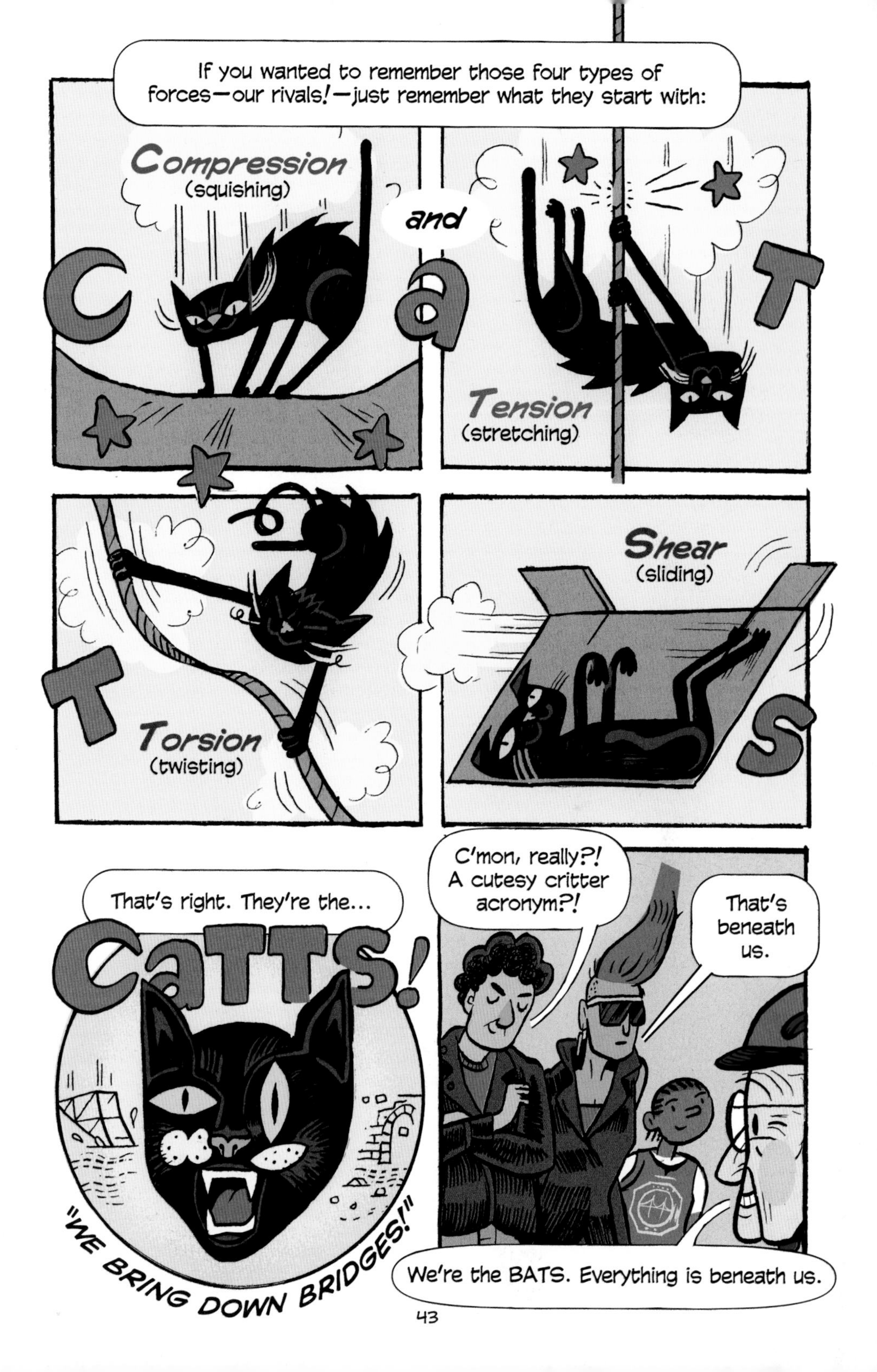

If you wanted to remember those four types of forces—our rivals!—just remember what they start with:
Compression
(squishing)
C
and
a
T
Tension
(stretching)
T
Torsion
(twisting)
Shear
(sliding)
S
That's right. They're the...
CaTTS!
"WE BRING DOWN BRIDGES!"
C'mon, really?! A cutesy critter acronym?!
That's beneath us.
We're the BATS. Everything is beneath us.

Structural engineers use a kind of "X-ray vision" to see how forces impact the bones of built structures.
Luckily for us, bridges have fully visible skeletons. We don't have to peel away layers of walls and stuff to see how they work. But it does take training to "see" the invisible CaTTS at work.
What do you mean "invisible"? I see them all! Aw...kitties.
Just kidding.
WINK!
Okay...my turn!
Let's look at the *best* kind of bridges...

SECTION
2
ARCH
BRIDGES
Get ahead of the curve.
Also!
Materials!
What are bridges made of?
Bonus!
Stress & Strain

Near Argolis, Greece
The first arch bridge we visit has been here since the Bronze Age!
ARKADIKO BRIDGE
ONE of the OLDEST BRIDGES STILL STANDING
Original bridge traffic: Pedestrians, chariots
Material: Limestone boulders piled up. The rocks are so huge they are known as Cyclopean stones.
Supposedly only the mythical giant Cyclops was able to lift 'em.

Ancient Greek soldiers passed over this highway bridge...
...and it's even still used today!
Type: **Corbel arch bridge**
Material: **Limestone**
Total length: ≈ **22 m**
Longest span: ≈ **1 m**
Opened: **c. 1300 BCE**
Forces on this bridge have been balanced for hundreds of years.
(These arrows show how invisible forces move through the structure.)
C'mon, Archie, go under with me! It's like a cave in here!
No thanks. I'll stay out here, where I can, uh, admire the arch.
BATS

Think it's gonna fall down?
No way! Arches are one of the strongest structures we have.
But I love them for their classical beauty.
KISS

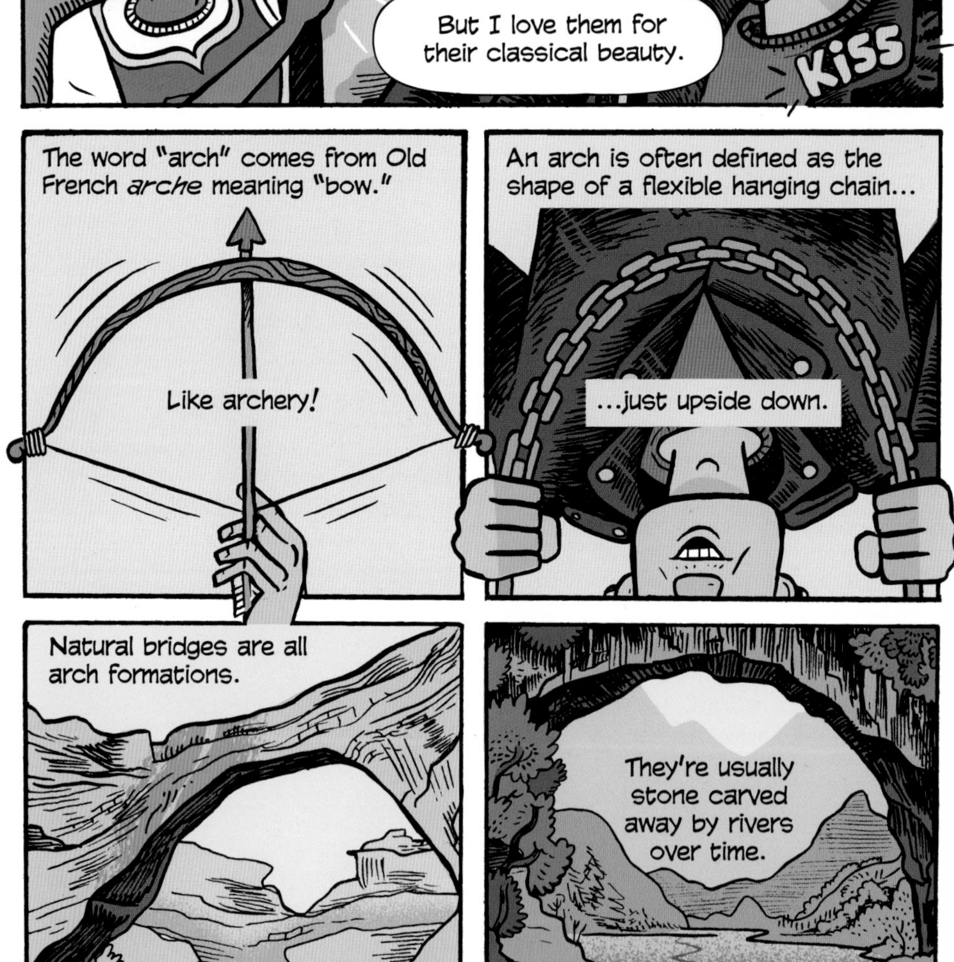
The word "arch" comes from Old French *arche* meaning "bow."
Like archery!
An arch is often defined as the shape of a flexible hanging chain...
...just upside down.
Natural bridges are all arch formations.
Sipapu Bridge, San Juan County, Utah, USA
They're usually stone carved away by rivers over time.
Xianren "Fairy" Bridge, Guangxi Province, China

There are two main types of arches:
Corbeled Arch
Uses blocks stacked up like stairsteps (like the Arkadiko Bridge).
True Arch
KEYSTONE
VOUSSOIR
ABUTMENT
Uses angled block wedges to create the more familiar round shape.
In our arch research, we'll focus on the STUFF they're made of. You'll see how bridge construction materials have evolved.
We'll examine how they stand up against their archenemies:
Loads and forces!
Actually, arches have just one force to deal with: compression. The force flows around the curve and pushes outward toward the bottom.
That's why you need a solid abutment.
A solid what now?

Segovia, Spain
A thousand years later, the Roman Empire was building bridges with not just one arch but hundreds!
The AQUEDUCT OF SEGOVIA
Bridge traffic: H_2O
This bridge isn't for people, horses, trains, or cars. Just water!
This aqueduct channeled water from the mountains all the way into the city.
It's an incredible feat of engineering and construction.
Too bad our incredible feet can't walk across it!
Material: Granite blocks fitted together precisely but with little or no mortar (paste) between

Check out the long-lasting power of these ARCHES!

Type: **Multi-arch aqueduct**
Material: **Granite blocks**
Aboveground length: **728 m**
Longest span: **5 m**
Height: **28.5 m**
Opened: **c. 112 CE**

180°

Romans used lots of semicircular arches in sequence, which distribute the compression forces nicely. *Arches experience primarily compression*.

Stone is very strong when it comes to compression, so it works great for arches.

Usually we think of bridges carrying people over water, not vice versa.

I guess *agua* was as important back then as it is now.

B.A.T.S.

BATS reusable water bottles available now!

Zhao County, Hebei Province, China
Adiós, Spain! We're back in China, crossing the Xiao River.
The ANJI BRIDGE
安济桥
Literally translated: "Safe crossing bridge."
The only thing scary about this bridge are these carved dragons!
BATS
Material: Smooth limestone slabs fitted together with mortar and iron joints

Type: Segmental arch bridge
Material: Limestone slabs
Opened: 605 CE (Sui Dynasty)
Total length: 50.8 m
Longest span: 37.4 m
Height: 7.23 m
Rather than a semicircular arch, the Anji Bridge uses a smaller segment of a circle for its curve.
90°
It also introduces these clever arched openings called spandrels, which allow high waters through.
They also allow the bridge to use less material and feel more graceful.
"It's a crescent moon rising from the clouds."
How poetic. Did you make that up yourself?
I read it on the plaque next to the bridge.

Trudy, pass me that wooden ruler again. And the microscope.
Let's really eyeball *materials* and how we measure their strength:
Stress is the *amount* of force pushing or pulling on an area of material.
Stress = Load / Area
Strain is how *far* a material is deformed due to the stress.
Strain = Length Increase / Original Length
Strength is the amount of stress needed to break a material.
CRACK!
Everything has its breaking point.

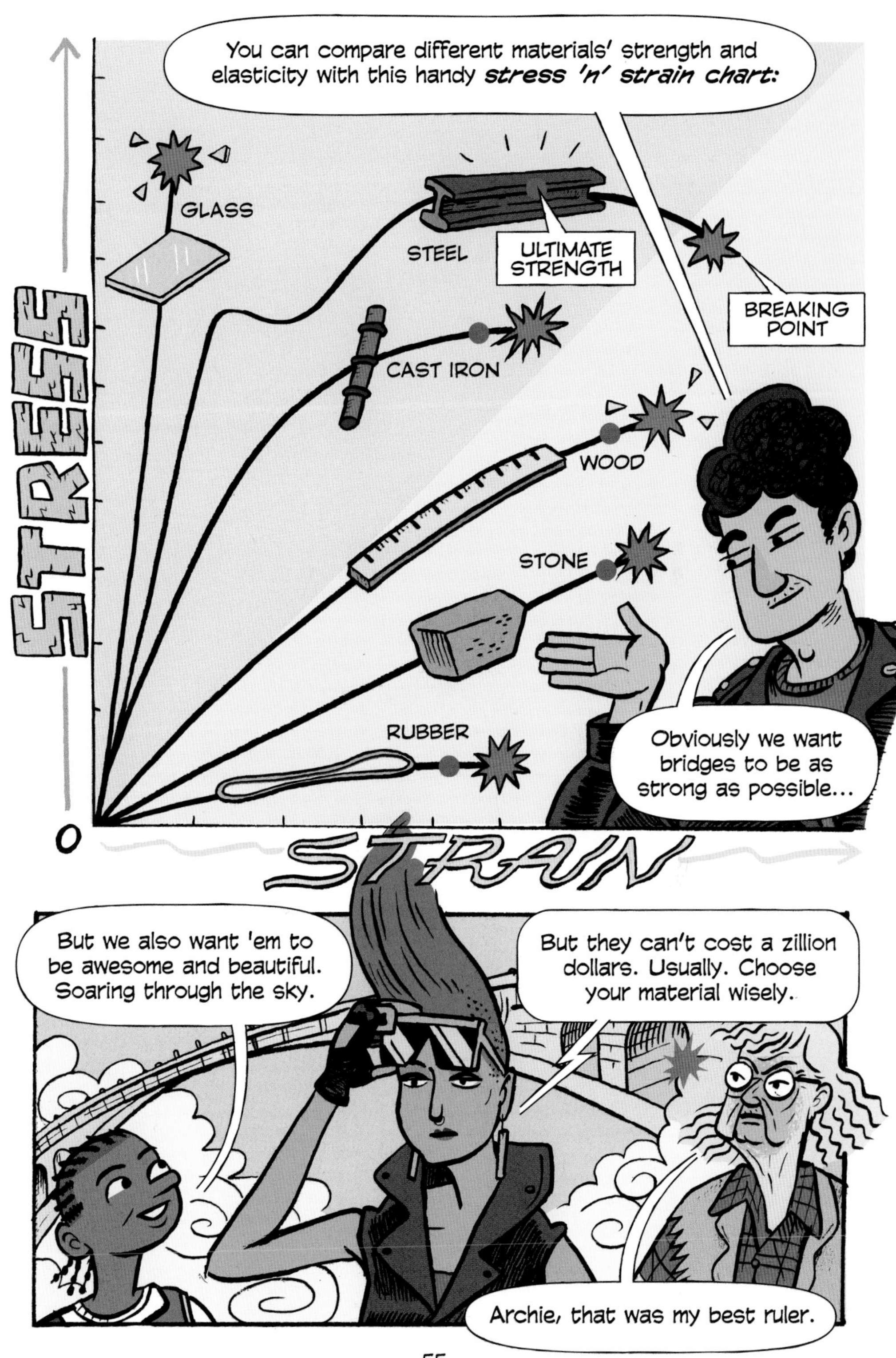
You can compare different materials' strength and elasticity with this handy ***stress 'n' strain chart:***
STRESS
GLASS
STEEL
ULTIMATE STRENGTH
BREAKING POINT
CAST IRON
WOOD
STONE
RUBBER
0
STRAIN
Obviously we want bridges to be as strong as possible...
But we also want 'em to be awesome and beautiful. Soaring through the sky.
But they can't cost a zillion dollars. Usually. Choose your material wisely.
Archie, that was my best ruler.

Isfahan, Iran
Moving to the Middle East, check out these fancy digs over the Zayandeh River.
The Khaju BRIDGE
This bridge has everything:
• two stories
• lots of arches
• a dam
• pleasant walkways
• tea shops
• hang-out zones
I need a rest.
Arch material: Persian brick (blocks of mud dried in the sun)
Decorative material: Glazed ceramic tiles
Base material: Stone slabs glued together with mortar

Type: **Multi-arch bridge**
Materials: **Stone, brick, and ceramic**
Total length: **137 m** Height: ≈ **20 m**
Opened: **1650**
Arches in the Islamic style are pointed.
Y'all remember the BATS motto, right?
Z z
"WE HANG AROUND BRIDGES!"

Shropshire, England, UK
First we had raw stone. Then precisely shaped rock. Then mortared brick. Now we've reached the Industrial Revolution, here at the River Severn.
IRON
Dang.
Dovetail joints inspired by woodworking
Abutment material: Stone
Bridge material: Hundreds of cast-iron metal parts, including the 5 giant ribs, weighing 384 tons

BRIDGE
Type: Single arch
Material: Cast iron
Total length: 60 m
Longest span: 30.6 m
Max clearance: 18 m
Opened: 1781
Cast iron is strong in compression.
FIRST MAJOR CAST-IRON BRIDGE
That's a lot of metal.
They don't call it the Iron Age for nothin'.
B.A.T.S.
B.A.T.S.

Sticking to this section's subtheme of *materials*, here's a short history of
HEAVY METALS
Iron is one of Earth's most common elements.
Iron ore is mined, crushed, then sent to the *blast furnace.*
IRON ORE
HOT AIR blasted in
COKE (fuel, not soda)
LIMESTONE
The result from the blast furnace is valuable *pig iron.*
Pig iron gets turned into different usable metals.
BESSAMER PROCESS

Cast iron is an alloy with a lot of carbon mixed in. It's a hard, brittle metal that is strong in compression.
Which is why it's great for an arch design.
(This bridge carries boats!)
Pontcysyllte Aqueduct, Wales
Wrought iron is an alloy with very little carbon. It's tough but more elastic and works better in tension.
Gustave Eiffel loved this stuff.
Maria Pia Bridge, Portugal
The Eiffel Tower, France
Steel is in between. It's strong but also flexible.
Remember the Eads Bridge in St. Louis? Back on pages 2–4?
Oh yeah, the genius elephant.

After steel, the next major revolution in materials came with *concrete*. Basic concrete has been around since ancient times. This Mesopotamian bridge was built in 690 BCE using waterproof cement.
The Jerwan Aqueduct, Iraq
CONCRETE
needs three ingredients mixed together:
Water
Aggregate, which can be sand, gravel, crushed stone, etc.
Cement, which is usually a powder made from limestone.
Pour your concrete into the shape you need and let it harden.
BLOCK
FLAT
DECORATIVE/ OTHER

Concrete is everywhere. We've covered the world with it.
But we didn't know its true strength until this little trick...

REINFORCED CONCRETE
Pouring concrete around steel reinforcing bars (aka *rebar*) creates a material that handles loads and forces with the best of both worlds.
Concrete: strong in compression but weak in tension, cracking easily
Steel: strong in tension, helping concrete not to crack

The cave-like *Alvord Lake Bridge* in San Francisco, California, is America's oldest reinforced concrete bridge.
30
Now this is a good place for BATS to hang out!

Between the Western and Eastern Capes of South Africa
We're atop the highest bridge in Africa, 216 meters above the Bloukrans River.
I'm not gonna be atop much longer!
Uh...sure you wanna do this, Trudy?
FEAR IS TEMPORARY REGRET IS FOREVER
CLICK
30
Of course! This is the best way to see...
Bridge firmly anchored to valley wall abutments.
THE BLOUKRANS BRIDGE!
!
Home of the world's highest bridge bungee jump!
SPROING!

Bridge material: ***Prestressed concrete.*** Steel tendons inside make the concrete "pre-compressioned" to counteract the stresses that will occur when in use. This deck shouldn't crack under pressure.
Type: **Single-arch bridge**
Material: **Prestressed and reinforced concrete**
Total length: **451 m**
Arch span: **272 m**
Opened: **1983**
Whaddya think, Spence? You gonna jump next time?
Uh, no way.
I'm ***prestressed*** just thinking about it.
**You can't see them here, but BATS always wear seat belts.*

Dude. Congratulations. You're on your way to truly being a BAT with these terrible bridge puns.
Thanks for the *reinforcement.*
Don't *strain* yourself.
Have you ever heard the old riddle: "What thing is *more* scary when there is *less* of it?"
A bridge!
Don't worry. In this next section, there will be plenty of *it.*

SECTION 3
TRUSS
BRIDGES
Try angles!
ALSO!
Design! How are bridges shaped?
Bonus! Build DIY model bridges...
...for fun and profit!

Otsego County, New York, USA
The first truss bridge we visit has no bungee jump.
HYDE HALL
Bridge
OUR OLDEST COVERED BRIDGE
Now y'all are entering my world, the world of △ triangles! ▽
I'll enter anybody's world to get outta this rain!
This bridge has got you covered!
B.A.T.S.
30

Type: **Burr arch truss**
Material: **Wood**
Total length: ≈ **16 m**
Single span: ≈ **10 m**
Clearance: ≈ **2.5 m**
Opened: **1825**
Is that why they built covered bridges? To keep people dry?
Actually it was to keep *the bridge* dry!
Keeping the inner structure safe—all these crisscrossing wood beams—from bad weather makes the bridge last a lot longer.
CUTAWAY VIEW
And it makes them totally romantic.
Ugh.
LES + DAN
Design: The ***Burr arch truss,*** invented in 1804 by Theodore Burr, combines the strength of an arch with the stability of a truss design.

Trudy, what's with you and triangles?
Triangles are just terrific! They're the strongest geometric shape. Look at these toothpick structures:
Look how easily you can shear and deform the square.
You can't do that to a triangle.

The word "truss" comes from Old French for "a collection of things bundled together."

Truss structures were first used as roof supports.
KING POST ROOF TRUSS

Early timber truss bridges could be found in heavily forested countries like Switzerland.

Now there's an "Integrated Truss Structure" on the International Space Station.

The idea is that if one triangle is strong...
...then multiple triangles bundled in a truss will be super strong.
All while using less material to build it.

Check out this *Howe truss,* an early popular design:
Trusses are like beams in that they deal with both *compression* and *tension.*
TOP CHORD
DIAGONAL MEMBER
VERTICAL MEMBER
BOTTOM CHORD
They do a great job of balancing forces and distributing the load throughout the structure.
Let's look at some in action.

Our journey across the globe brings us to this small park outside Kansas City.

WADDELL "A" TRUSS BRIDGE

Type: **Waddell "A" Truss**
Material: **Steel**
Longest span: **30.48 m**
Height: **12.2 m**
Opened: **1898**
This bridge is so nice they built it twice.
(It was taken apart, moved from its old location, moved here, then reassembled.)
Okay, but are we gonna see any BIG truss bridges?
Ha! We'll get there, young man.

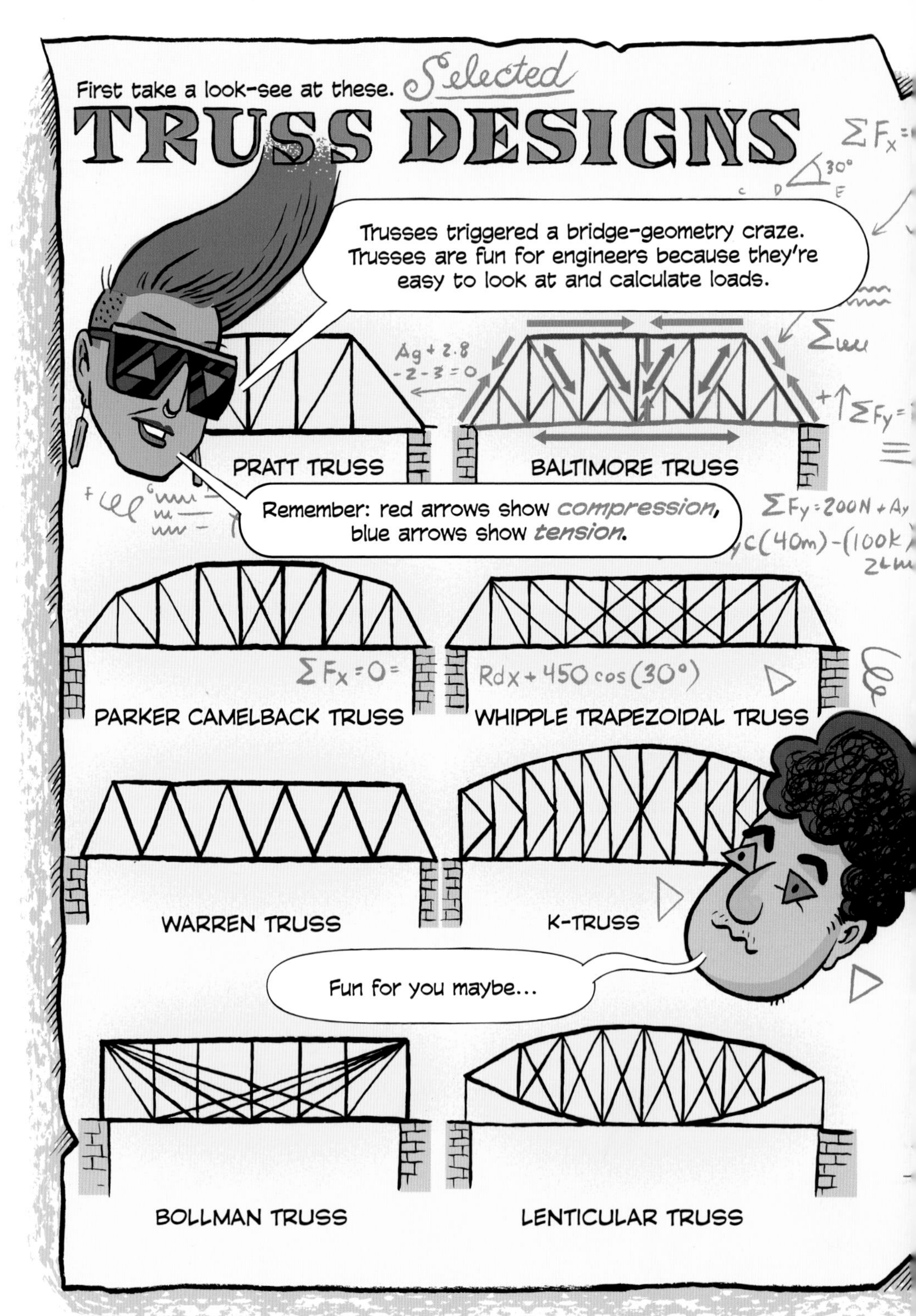
First take a look-see at these. Selected
TRUSS DESIGNS
Trusses triggered a bridge-geometry craze. Trusses are fun for engineers because they're easy to look at and calculate loads.
PRATT TRUSS
BALTIMORE TRUSS
Remember: red arrows show compression, blue arrows show tension.
PARKER CAMELBACK TRUSS
WHIPPLE TRAPEZOIDAL TRUSS
WARREN TRUSS
K-TRUSS
Fun for you maybe...
BOLLMAN TRUSS
LENTICULAR TRUSS

PONY TRUSS
DECK TRUSS
The previous trusses are considered "through trusses" (because you pass through them). A "pony truss" is only half-height.
Deck trusses look like they've been built upside down!
FINK TRUSS
WICHERT TRUSS
These were all attempts to maximize the strength of a bridge while minimizing the cost. Builders didn't want to spend so much money on wood, iron, or steel.
However, the builders of this next bridge spared no expense.
MULTI-SPAN CANTILEVER THROUGH TRUSS

Near Edinburgh, Scotland, UK
We're on the banks of the Firth of Forth, looking at the huge...
FORTH BRIDGE
THE STRONGEST BRIDGE IN THE WORLD!
• 49,000 tons of steel • 148,000 cubic meters of masonry • 7 million rivets
There ya go, Spence. Told ya we'd get to a big one!
I heard it's so huge that once they finish painting it it's time to start painting again.
Design: This gigantic steel truss features two suspended spans between the massive ***cantilever*** supports.

Here are the BATS demonstrating the ***cantilever*** principle with young Spence suspended between horizontal supports. Try this at home!

BUILD YOUR OWN BRIDGE!
"DIY!"
You don't need 49,000 tons of steel to build a bridge. Some of this household stuff will do the trick!
STRING
SPAGHETTI (OR FETTUCCINE)
GLUE
RUBBER BANDS
TOOTHPICKS
FISHING LINE
GUMMY CANDY
POPSICLE STICKS
Have a bridge-building competition with your friends. Win fame or prizes.
Or just do it for fun. And to eat the leftover noodles and popsicles.
SSSLLLURP
25 kg
50 kg
I prefer macaroni, by the way. They're shaped like little arches.

This girl is crushing it!
PHYSICS
You'll get a hands-on feeling for what structures are strongest and why.
PHYSICS
PHYSICS
This is gonna blow up on YouTube.
TRIG
SCIENCE
We're outta physics books! Go get some art history!
GEOMETRY
PHYSICS
PHYSICS
PHYSICS
Easy, easy...
No matter the reason, model bridge tests usually end with a...
CRASH!
Organize one for your school or neighborhood today!

Edmonton, Alberta, Canada
Here's another massive bridge, one that does it all.
The HIGH
Are we gonna take the streetcar? Walk? Drive the lower deck?
None of the above!
Bridge traffic: Four-way! Automobile, train, streetcar, pedestrian
There used to be a huge artificial waterfall that poured off of this bridge.
Water is the only good thing that can fall off a bridge.

LEVEL BRIDGE
Type: **Truss**
Materials: **Steel and concrete**
Total length: **777.24 m**
Longest span: **88 m**
Max clearance: **47.55 m**
Opened: **1913**
The view is the best from the upper railway deck.
Good thing we got the Echolocator fitted with road-rail wheels!
CLACK CLICK CLACK CLICK CL
Design: Double-decker bridge using mostly Pratt trusses but also a few Warren trusses.

Sometimes it's not just cars, trains, or pedestrians moving on bridges, it's the bridges themselves! Here are a few...
MOVABLE BRIDGES
You mean like drawbridges?
Yes! There are many kinds. No castle required.

The Tower Bridge, London, England, UK
BASCULE
B.R.T.S.

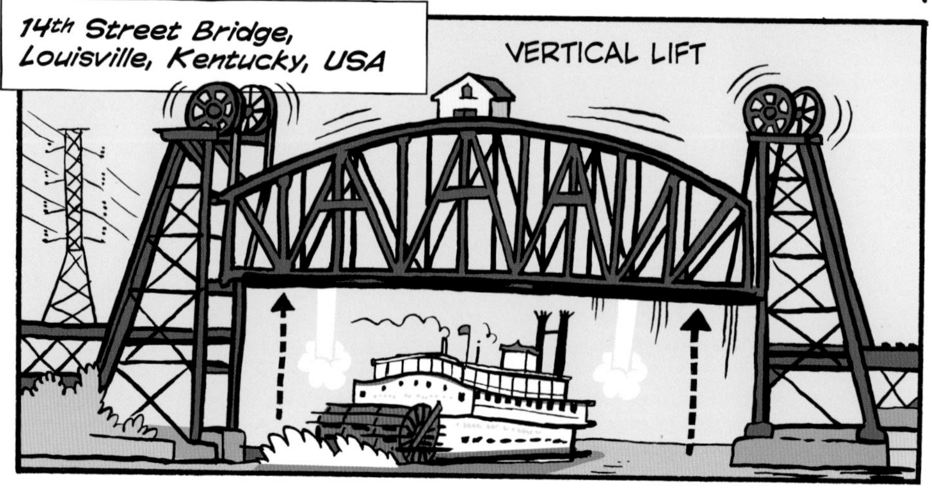
14th Street Bridge, Louisville, Kentucky, USA
VERTICAL LIFT

Puente de la Mujer, Buenos Aires, Argentina
SWING BRIDGE

Vizcaya Bridge, Biscay, Spain
TRANSPORTER BRIDGE
Okay...
This is officially the weirdest bridge so far.

Ganges River Bridge, Allahabad, India
PONTOON BRIDGE

Los Angeles, California, USA
Now we're in a city dominated by cars but we're crossing this small-scale truss by foot.
LA Pedestrian BRIDGE
And pedal!
Sometimes you just gotta enjoy a bridge at strolling speed.
Scenic.
B.A.T.S.
Design: Warren pony truss on wood and concrete piers.

Type: **Truss**
Materials: **Steel, wood, concrete, and spray paint**
Total length: **96.32 m**
Longest span: **14.6 m**
Deck width: **1.5 m**
Opened: **1929**
What's with all these junky padlocks all over it?
Romantic types like Archie think it's sweet to vandalize bridges with "love locks." Ugh.
Hey! Even I know that they're a structural hazard and can damage a bridge.
IN BRIDGES WE TRUSSED
Juliet
So keep your locks on your lockers or bikes.
And keep love in your hearts!

Kōtō City, Tokyo, Japan
TOKYO
We've crossed the Pacific Ocean (don't ask how) and have arrived at this gargantuan truss.
This was a tough one. Engineers had to make it low enough to not get in the way of planes.
But still with enough clearance for big ships.
And it's gotta stand up against earthquakes and other catastrophes.
Design: Double cantilever truss that appears to be two giant lizards face to face, giving it the nickname Dinosaur Bridge.
B.A.T.S.
B.A.T.S.

GATE BRIDGE

東京ゲートブリッジ

Type: **Cantilever truss**
Material: **Steel**
Total length: **2,618 m**
Height: **87.8 m**
Max clearance: **54.6 m**
Longest span: **440 m**
Opened: **2012**

As a bridge designer, that's what I call a *cough* mons***TROUS*** challenge.

Bea! Did you just make a ***truss*** pun?!

Of course not.

Spence, we've all discussed it. And we think you're ready.

Here ya go.

'Bout time! Fits great.
Your style will add a lot to the BATS.
WE HANG AROUND
BRIDGES
B A T S

For sure. My style's the best.
Not sure about that, dude.

Prepare to *suspend* your disbelief.

SECTION
4
It's hang time!
SUSPENSION
BRIDGES

ALSO!

Construction!
How are bridges built?

Bonus!
Cable-stayed bridges...

...and more!

Quehue District, Canas Province, Peru
We're in South America in the Andes Mountains, 3,700 meters above sea level, staring down this historic bridge.
Q'eshwachaka
THE OLDEST INCA ROPE BRIDGE!
Whoa, Spence! You're not afraid of this one? It's the scariest bridge yet!
Hmmm...I guess you're right!
B.A.T.S.
Construction method: This ancient bridge is rebuilt every June by residents in the traditional method:
1: Local grass is cut, dried, and twisted into cords.
2: The cords are braided into ropes that stretch across the gorge.

Type: **Simple suspension bridge**
Material: **Ichu grass fiber**
Total length / span: ≈ **30 m**
Clearance: ≈ **18 m**
Opened: **c. 1400 CE**
Maybe I'm not scared because I feel more at home on suspension bridges!
Or maybe it's just my new jacket.
3: Offerings are made to the mountain spirits and the old bridge is cut down.
4: Brave builders complete the handrails and sides.
5: A great festival celebrates the bridge's renewal.

Y'all know that *suspension* bridges have been around for a while, right?
Spiders have been building 'em for millions of years.
These webs are all tension and are really strong for how light they are.

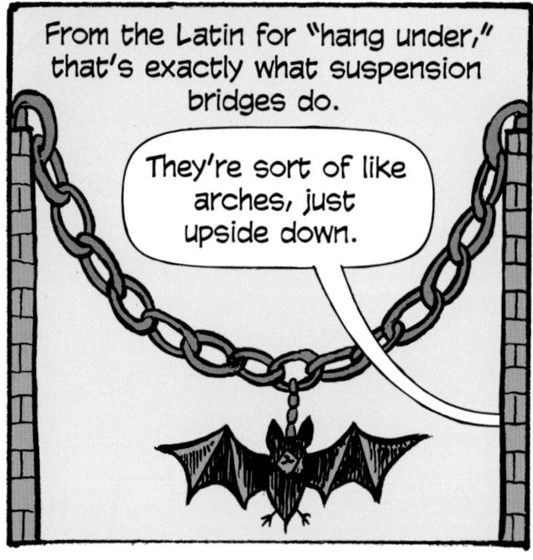

From the Latin for "hang under," that's exactly what suspension bridges do.
They're sort of like arches, just upside down.

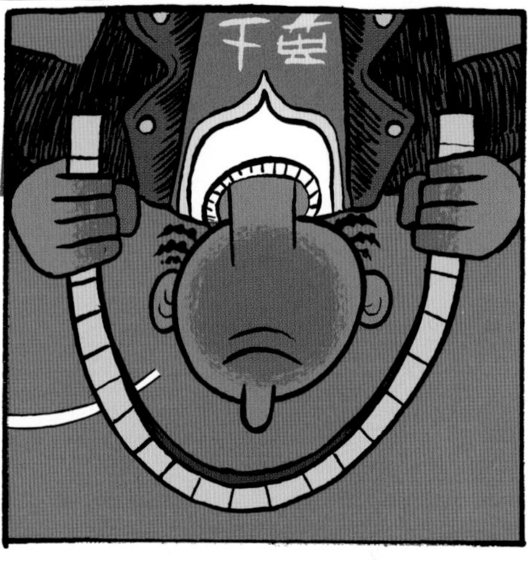

We often say bridges are "living things" but it's really true of India's
living root bridges.

This is taken to a new level by army ants who build suspension bridges out of...themselves!
Ew!

Look in my sketchbook. This shows the parts of a suspension bridge. Lemme draw in how they balance *compression* and *tension*.

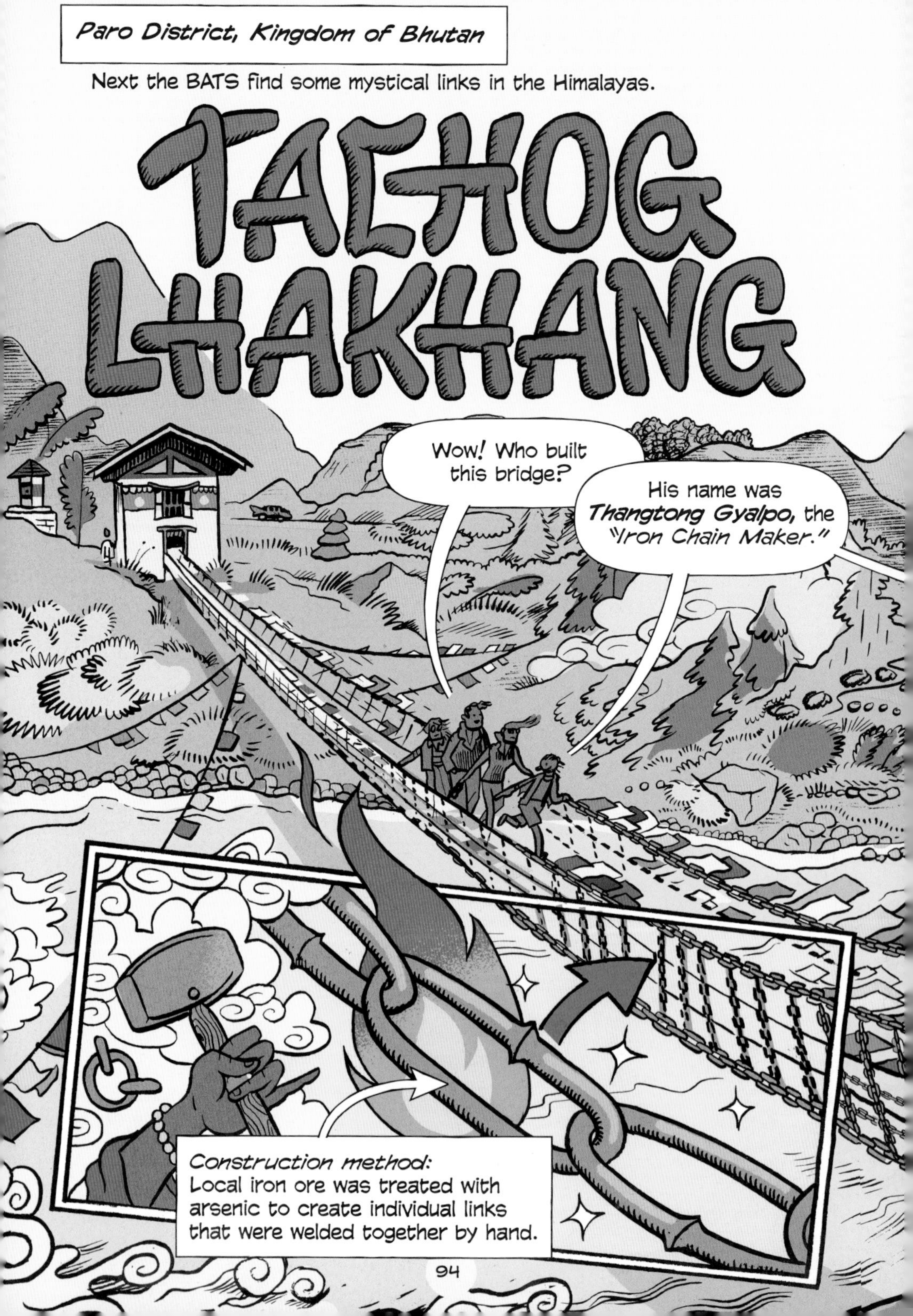

Paro District, Kingdom of Bhutan

Next the BATS find some mystical links in the Himalayas.

TACHOG LHAKHANG

Wow! Who built this bridge?

His name was ***Thangtong Gyalpo,*** the *"Iron Chain Maker."*

Construction method:
Local iron ore was treated with arsenic to create individual links that were welded together by hand.

Type: **Simple suspension bridge**
Material: **Iron chain links**
Total length / span: ≈ **27 m**
Clearance: < **5 m**
Opened: **c. 1430 CE**
Rebuilt: **2005**
He built more than 50 suspension bridges throughout Bhutan and Tibet, but was more than just an engineer and blacksmith. He was a true celebrity!
But he might've been better known as the *Madman of the Empty Valley*!
How many bridge-builders can you say are actual celebrities?
More than you think!

Scientists study the natural world through experiments and observations.

Engineers apply that knowledge and create the BUILT environment.

ENGINEERS!

CIVIL ENGINEERS
This broad category, distinguished from military engineers, goes way back.

We're in charge of large projects like:

AIRPORTS

ROADS

RAIL TRANSPORT

WATER SYSTEMS AND SEWERS

ETC.

STRUCTURAL ENGINEERS
This subcategory of builders specializes in load-bearing structures.

We focus on building and maintaining:

BUILDINGS

ROLLER COASTERS

BRIDGES

DAMS

ETC.

Thangtong Gyalpo was also a Buddhist master, yogi, doctor, architect, and theater troupe leader!

Hey, I sing a little bit here and there!

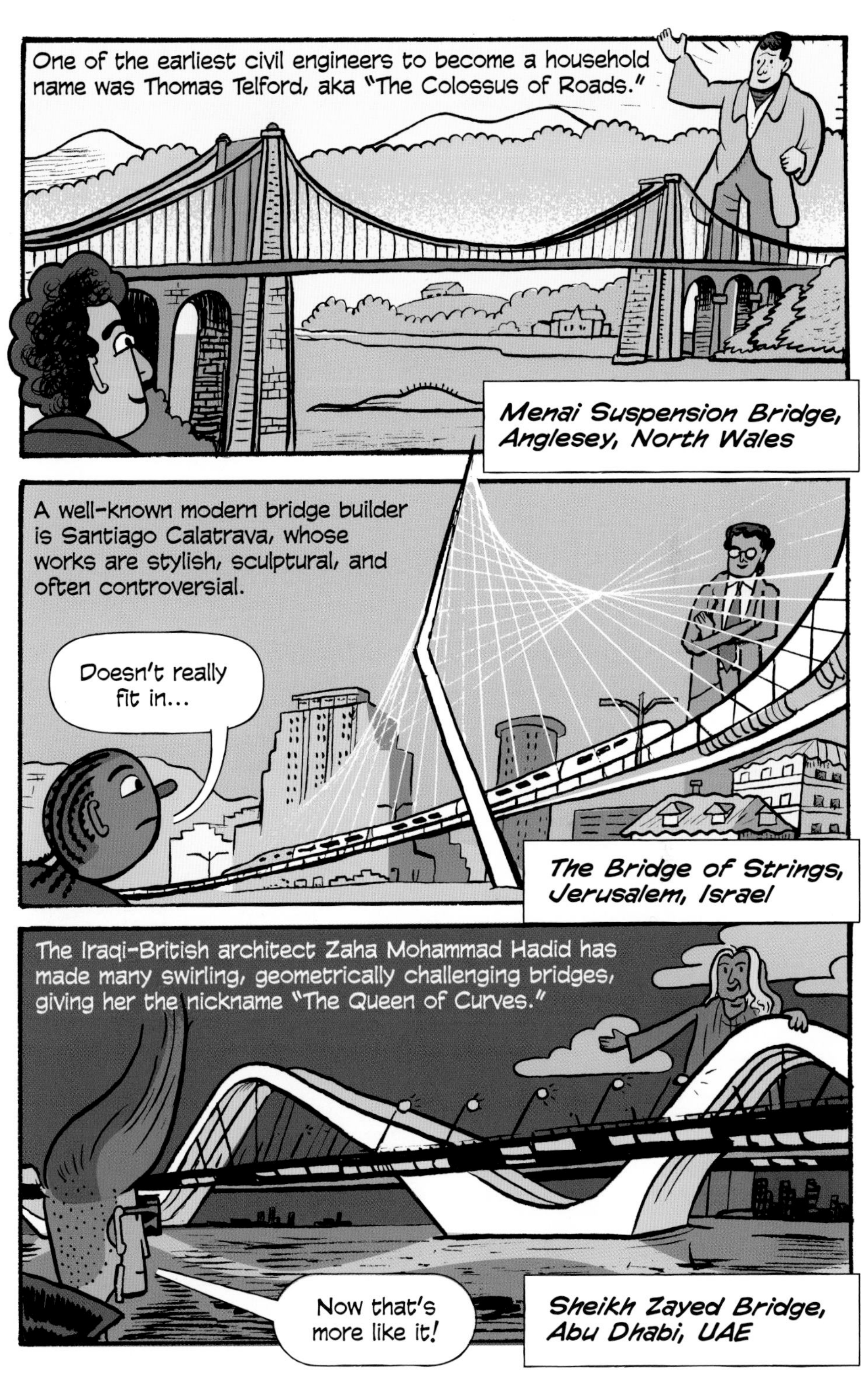
One of the earliest civil engineers to become a household name was Thomas Telford, aka "The Colossus of Roads."
Menai Suspension Bridge, Anglesey, North Wales
A well-known modern bridge builder is Santiago Calatrava, whose works are stylish, sculptural, and often controversial.
Doesn't really fit in...
The Bridge of Strings, Jerusalem, Israel
The Iraqi-British architect Zaha Mohammad Hadid has made many swirling, geometrically challenging bridges, giving her the nickname "The Queen of Curves."
Now that's more like it!
Sheikh Zayed Bridge, Abu Dhabi, UAE

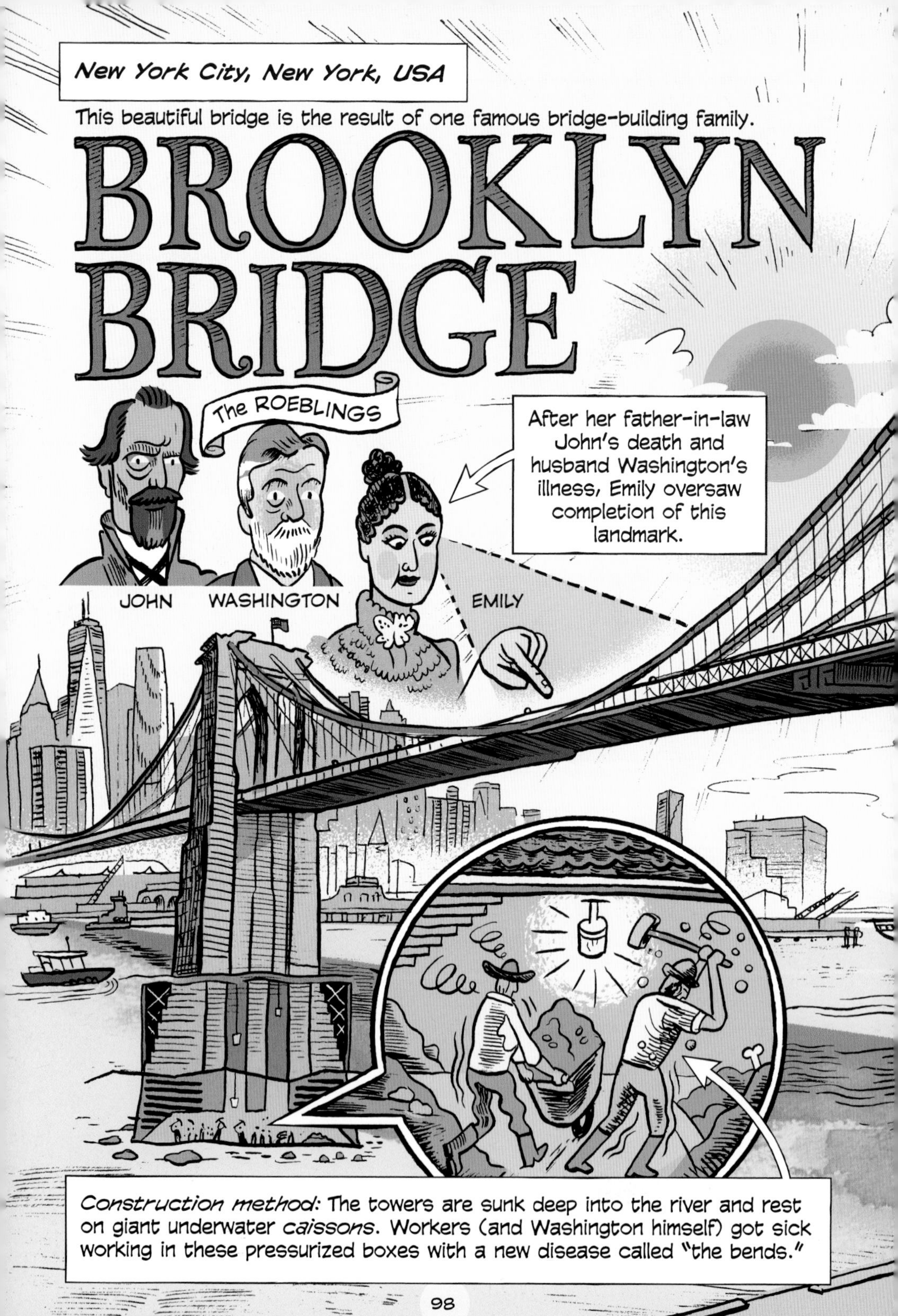

New York City, New York, USA

This beautiful bridge is the result of one famous bridge-building family.

BROOKLYN BRIDGE

The ROEBLINGS

After her father-in-law John's death and husband Washington's illness, Emily oversaw completion of this landmark.

JOHN

WASHINGTON

EMILY

Construction method: The towers are sunk deep into the river and rest on giant underwater *caissons*. Workers (and Washington himself) got sick working in these pressurized boxes with a new disease called "the bends."

Type: **Suspension / cable-stayed bridge**
Materials: **Steel and masonry**
Total length: **1,834 m**
Longest span: **486 m**
Height: **83 m**
Clearance: **38.7 m**
Opened: **1883**
Dudes. Are we gonna cross it or what?
Lemme finish sketchin' it first. It's the perfect combo of science and art.
Brooklyn BRIDGE
Have you been drawing all these bridges this whole time?!

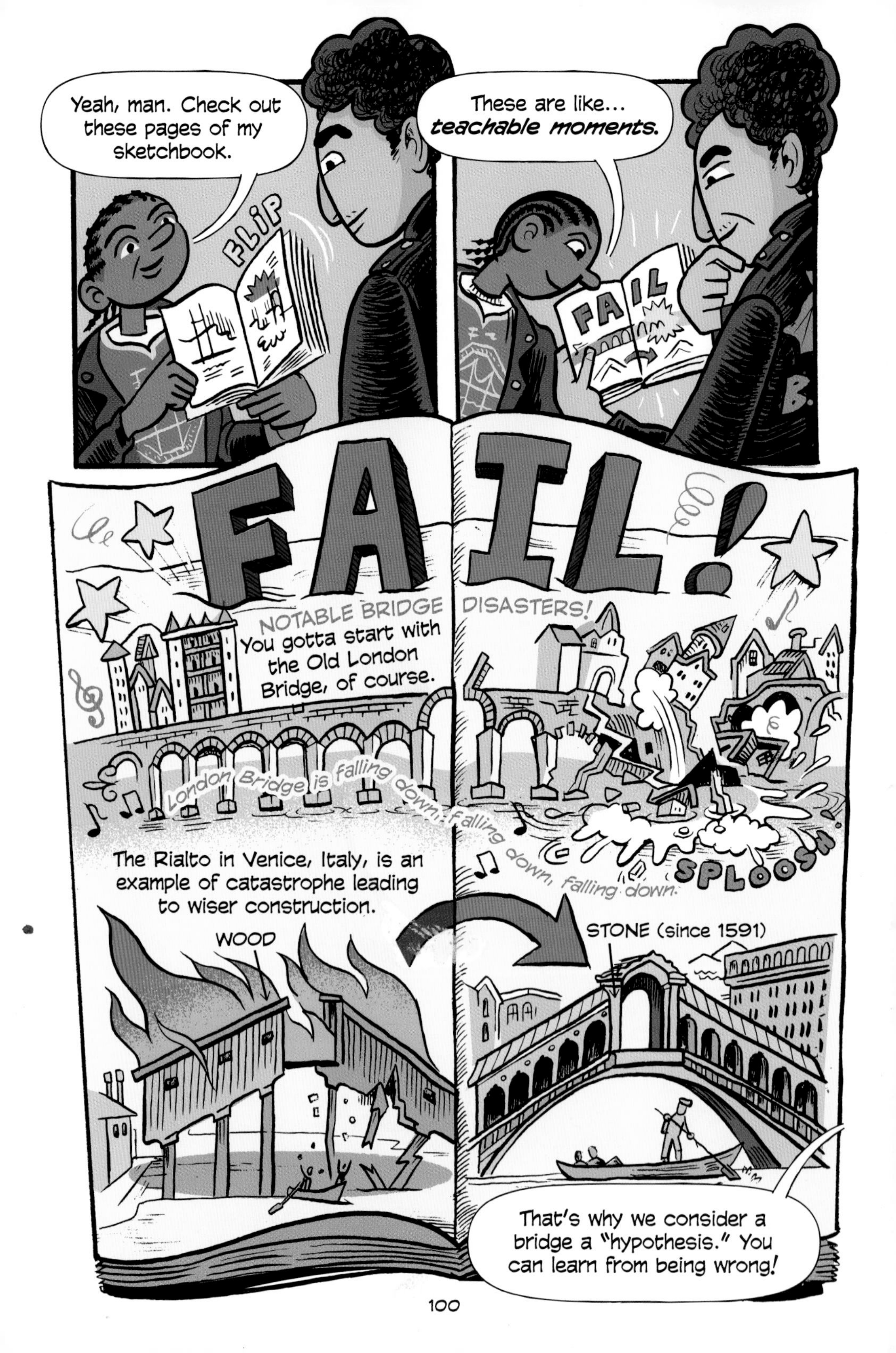
Yeah, man. Check out these pages of my sketchbook.
FLIP
These are like... *teachable moments.*
FAIL
FAIL!
NOTABLE BRIDGE DISASTERS!
You gotta start with the Old London Bridge, of course.
London Bridge is falling down, falling down, falling down.
SPLOOSH!
The Rialto in Venice, Italy, is an example of catastrophe leading to wiser construction.
WOOD
STONE (since 1591)
That's why we consider a bridge a "hypothesis." You can learn from being wrong!

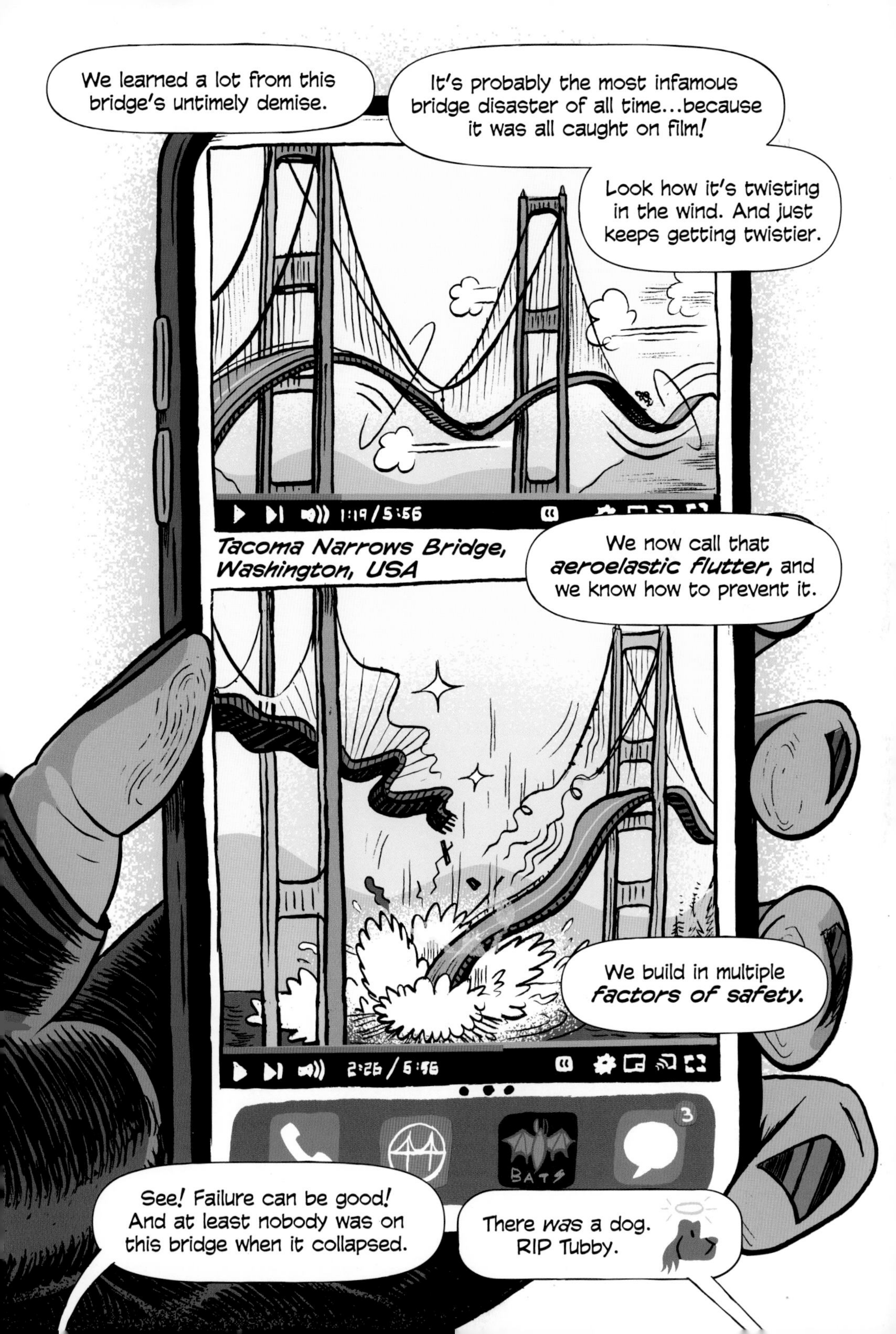
We learned a lot from this bridge's untimely demise.
It's probably the most infamous bridge disaster of all time...because it was all caught on film!
Look how it's twisting in the wind. And just keeps getting twistier.
1:19/5:56
Tacoma Narrows Bridge, Washington, USA
We now call that *aeroelastic flutter,* and we know how to prevent it.
We build in multiple *factors of safety.*
2:26/5:56
BATS
3
See! Failure can be good! And at least nobody was on this bridge when it collapsed.
There *was* a dog. RIP Tubby.

San Francisco, California, USA
The BATS arrive at the world's most iconic crossing.
GOLDEN GATE
When built, it was the longest and tallest suspension bridge in history!
Building this bridge was dangerous! Especially for the high-wire workers. New safety innovations included:
HARD HATS
GLARE-FREE GOGGLES
SAFETY ROPES
CURES FOR DIZZINESS
A GIANT SAFETY NET
Construction method: Each of the bridge's giant cables is made up of thousands of individual steel wires bound into strands then bound together.

BRIDGE
Type: **Suspension with truss arch approach**
Material: **Steel (international orange)**
Total length: **2,737 m**
Longest span: **1,280 m**
Height: **227 m**
Clearance: **67 m**
Opened: **1937**
Is it the most beautiful in the world?
Hmmm...hard to say. After all, beauty is in the eye of the bridge ogler.
I do think we can safely say it's the most photographed bridge in the world.
CLICK!
BATS selfie! Say beeeeeam!

I mentioned them earlier, but let's take a closer look at a few

CABLE-STAYED BRIDGES!

They look similar to suspension bridges but are actually quite different.

Cable-stayed bridges don't have large, curved suspension cables joining the towers with smaller cables hanging down. They have straight cables directly connecting the bridge deck to the towers.
Cavenagh Bridge, Singapore
The cables have a triangular, fanlike design or a series of parallel diagonal lines.
Most SNP, aka UFO Bridge, Bratislava, Slovakia

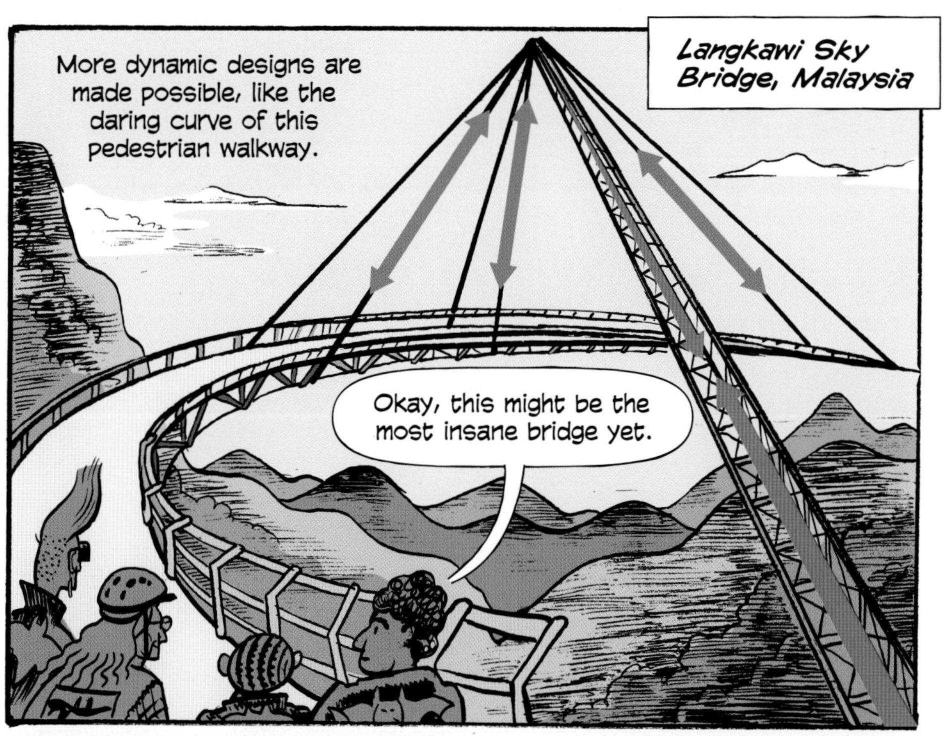
Langkawi Sky Bridge, Malaysia
More dynamic designs are made possible, like the daring curve of this pedestrian walkway.
Okay, this might be the most insane bridge yet.

Here's a bridge we can all love, as it's a hybrid of everything we've seen so far!
Maybe someday that hanging restaurant will open and it can be BATS headquarters.
Zhivopisny Bridge, Moscow, Russia

Tarn Valley, Millau, France
The BATS cross a truly epic cable-stayed bridge.
MILLAU
Whoa. Very *high tech.*
Construction method:
The road deck was installed high atop the piers with pinpoint precision. It was slid along smoothly by hydraulic machinery and met up with the other side only off by a few millimeters.

WORLD'S TALLEST BRIDGE!
VIADUCT
Type: **Multi-span cable-stayed**
Materials: **Reinforced concrete / steel**
Total length: **2,460 m**
Longest span: **342 m**
Height: **343 m**
Clearance: **270 m**
Opened: **2004**
Modern bridges rely on computers at every stage: research, design, modeling and simulation, construction...
And to think, the old-timers only had their slide rule!

Guizhou and Yunnan Provinces, China
The BATS reach the last bridge in this year's expedition, an intimidating cable-stayed beast.
It's like the final boss.
Construction method: China has been going through a bridge-building explosion. Most of the world's highest bridges call this mountainous region home. Extreme geological conditions force bridges higher and higher.

DUGE BEIPANJIANG BRIDGE
杭瑞高速北盘江大桥
WORLD'S HIGHEST* CROSSING!
Type: Cable-stayed
Materials: Reinforced concrete / steel
Total length: 1,340 m
Longest span: 720 m
Height: 269 m
Clearance: 565 m
Opened: 2016
565 m
EMPIRE STATE BUILDING
GREAT PYRAMID OF GIZA
I'll take this over the Great Wall any day.
Instead of dividing people, it connects us!
*Highest refers to the distance from the deck down to the ground or water below, i.e., clearance.
Tallest measures any part of the physical structure, i.e., the Millau Viaduct's towers.

Which brings us to the...
FUTURE
Beyond just higher, taller, and longer...
bridges keep reaching for unexplored horizons.
NEW PURPOSES
NEW MATERIALS
Green spaces
Fiber-reinforced plastics
Self-healing concrete
BACTERIA!
Community gathering
Visionary! BATS aren't actually blind, y'know.
Art experiences
Vehicle-free travel
B.A.T.
B.A.T.S.

BRIDGES!
NEW DESIGNS
NEW METHODS
Accessible
Environmentally sustainable
3D-printed components
Crazy unfolding ocean-cruising bridge?
Future bridges will need all these new ideas.
Most of all they'll need new engineers...
Maybe you?!
B.A.T.S.
BRIDGES

Well, BATS, we've reached the end of another adventure.
SPENZ
Here's to seeing the world's greatest connections...
...and making new ones!
3, 2, 1...
BRiDGES!

Remember, keep your Bridge Logs updated.
BRIDGE LOG
And 'til next time...
Good luck with life's ups and downs!

BASIC BRIDGE

Sydney Harbor Bridge, New South Wales, Australia

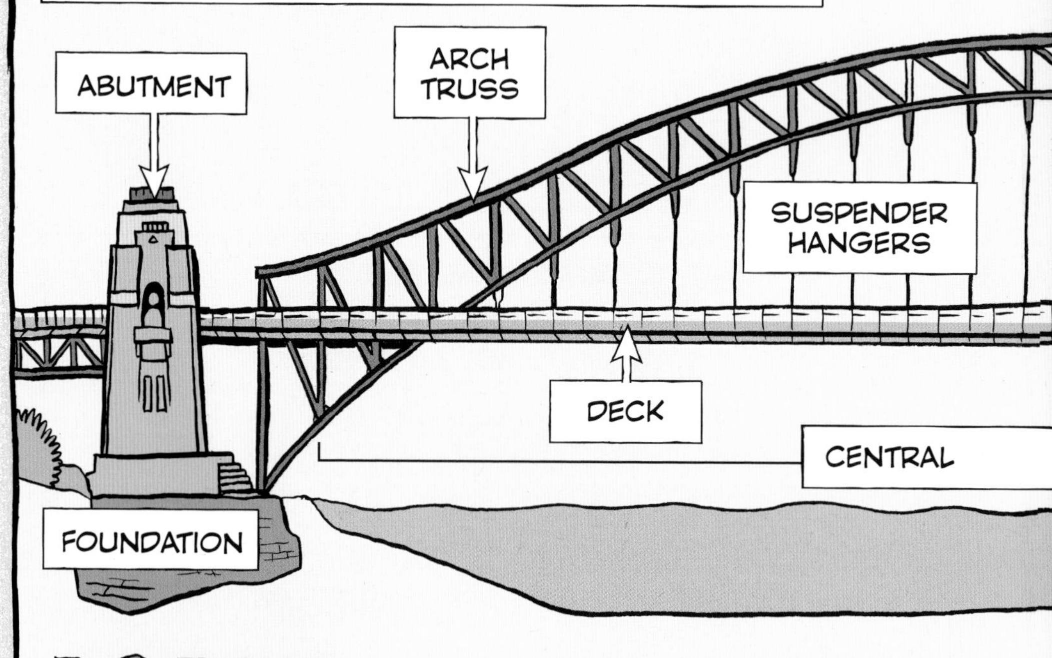

LOADS

LIVE LOAD

ENVIRONMENTAL LOAD

DEAD LOAD

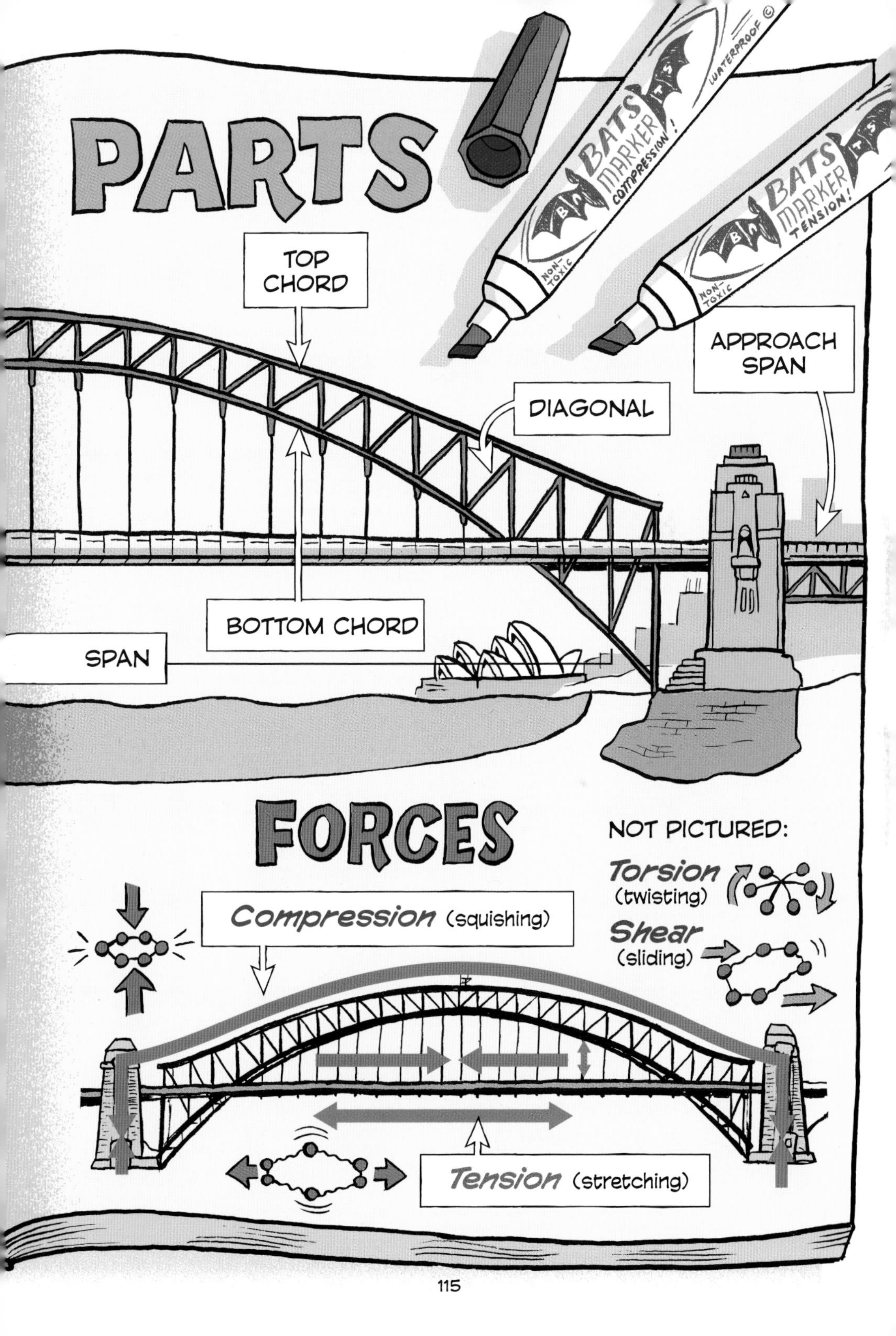
PARTS
BATS MARKER
COMPRESSION!
WATERPROOF ©
NON-TOXIC
BATS MARKER
TENSION!
NON-TOXIC
TOP CHORD
APPROACH SPAN
DIAGONAL
BOTTOM CHORD
SPAN
FORCES
NOT PICTURED:
Torsion (twisting)
Shear (sliding)
Compression (squishing)
Tension (stretching)

GLOSSARY

Abutment a structure supporting the end of a bridge, especially an arch

Aggregate the sand, gravel, or stone mixed with cement to form concrete

Aqueduct a bridge that carries water

Arch a curved structure spanning an opening, mostly under compression

Bascule a movable bridge that rotates from a horizontal to a vertical position; French for "see-saw"

Beam a straight structure spanning one support to another, experiencing compression and tension

Bioengineering the use of biological organisms or processes in design or construction

Blast furnace an industrial tower that uses hot compressed air to turn iron ore into usable metals

Box girder a hollow structural beam, typically with a square, rectangular, or trapezoidal cross-section

Breaking point the moment of greatest strain before a material fractures

Cable a collection of strands or a bundle of parallel wires forming a thick rope

Cable-stayed bridge a bridge in which the deck is directly supported by cables attached to towers, usually in a fanlike or diagonal pattern

Caisson a watertight chamber used while constructing bridge foundations in water

Cantilever a structure that extends freely beyond its supporting pier

Cast iron iron/carbon alloy that is hard, brittle, and strong in compression

Causeway a long, raised road over low or wet ground

Cement a powder that mixes with water and aggregate to form concrete

Chord the outer parts of a truss structure that defines its shape

Clapper bridge an ancient type of beam bridge made from piles of stones

Compression a "squishing" force that pushes a material together and makes it shorter

Concrete a mixture of aggregate, water, and cement that hardens into a stonelike material

Corbeled arch an opening formed by stacks of blocks gradually pushing inward

Covered bridge a bridge protected from environmental loads with a roof

Dead load the weight of the bridge material itself

Deck the surface part of a bridge that is a road, rail, or walkway

Deform to change the physical shape of a material through force

Drawbridge *see* movable bridge

Elasticity the ability of a material to spring back to its original shape after a deforming force is released

Engineer a person who designs, builds, and maintains structures, machines, or public works

Environmental load a dynamic load, often weather-related

Factor of safety a margin built into the hypothesized strength of a structure

Force an influence that pulls, pushes, or changes direction of an object

Foundation the lowest load-supporting part of a bridge

Gephyrophobia the fear of bridges

Gravity the natural phenomenon that pulls structures toward the earth

Hypothesis a proposed explanation that must be tested

I-beam a structural member with an I-shaped cross-section

Iron ore a raw metallic element used in cast or wrought iron and steel

Keystone the wedge-like piece placed in the center of an arch

Live load a dynamic load such as vehicular or pedestrian traffic

Load the weight carried by a structure

Masonry stone, brick, or concrete block construction

Materials the physical matter from which a bridge is made

Mortar a paste of cement, sand, and water used in masonry

Movable bridge a bridge with moving parts that can transform to allow through passage

Natural bridge a rock formation where an arch has formed

Pedestrian bridge a bridge used by walkers and often cyclists

Pier a column that supports the spans of a bridge

Pontoon bridge a bridge that floats on the water's surface

Pony truss a short through truss with an open top

Prestressing applying forces to a material to deform it, making it stronger during use

Reinforced concrete concrete with steel bars inside to increase strength and durability

Rust a type of corrosion that's damaging to iron and steel

Segmental arch a curved segment of an arch that's less than a semicircle

Shear a "sliding" force parallel to the surface of a material

Slide rule a mechanical engineer's tool before calculators or computers

Span the distance between bridge supports

Steel strong, durable iron/carbon alloy

Strain how far a material is deformed due to stress

Strand many wires twisted together

Strength the amount of stress needed to break a material

Stress the amount of force pushing or pulling on an area of material

Substructure the lower part of a bridge (i.e., abutments and piers) that supports the superstructure

Superstructure the upper part of a bridge (i.e., deck and towers) that carries traffic

Suspender hanger a vertical cable, rod, or bar connecting the main cable of a suspension bridge to the deck

Suspension bridge a bridge in which the deck is supported by hangers attached to a main cable strung between towers

Tension a "stretching" force that pulls a material apart and makes it longer

Torsion a "twisting" force

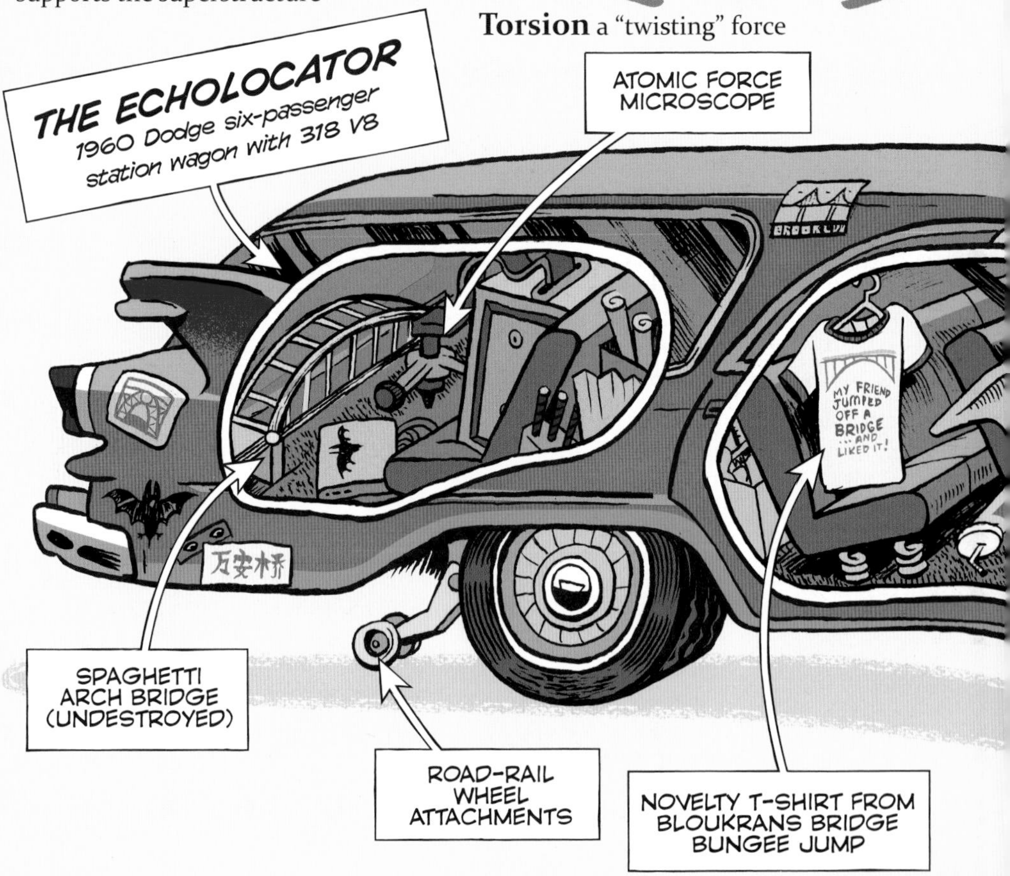

Tower a tall structure supporting the main cables of a suspension bridge or cable-stayed bridge

Trestle a framework bridge built on many pairs of angled legs

True arch *see* arch

Truss a structure of many parts arranged and connected together, usually in triangles

Viaduct a bridge made from many small spans

Voussoired arch an opening formed by a smooth curve of wedge-shaped blocks

Wrought iron iron alloy that contains little carbon and is softer than cast iron or steel

BIBLIOGRAPHY

Agrawal, Roma. *Built: The Hidden Stories Behind Our Structures.* Bloomsbury USA, 2018.

Blockley, David. *Bridges: The Science and Art of the World's Most Inspiring Structures.* Oxford University Press, 2012.

Dupré, Judith. *Bridges: A History of the World's Most Spectacular Spans.* Black Dog & Leventhal, 2017.

Gordon, J.E. *Structures: Or Why Things Don't Fall Down.* Da Capo Press, 2003.

Isaac, P.M. *A Critical Analysis of the Bloukrans Bridge.* University of Bath, 2007.

Karas, Slawomir and Nien-Tsu Tuan. "The World's Oldest Bridges - Mycenaean Bridges." *American Journal of Civil Engineering and Architecture,* 2017, Vol. 5, No. 6, 237–244.

Loomis, Richard Thomas. *The History of the Building of the Golden Gate Bridge.* Stanford University, 1958.

Mars, Roman and Emmett FitzGerald. "Rebar and the Alvord Lake Bridge" and "The Batman and the Bridge Builder." 99% Invisible. Podcast audio episodes #81 and #432, 2013 and 2021.

McCullough, David. *The Great Bridge: The Epic Story of the Building of the Brooklyn Bridge.* Simon & Schuster, 1983.

Miller, Howard and Quinta Scott. *The Eads Bridge.* Missouri History Museum Press, 1999.

Petroski, Henry. *To Engineer Is Human: The Role of Failure in Successful Design.* Vintage, 1992.

Sevaistre, Bruno. *The Bridge in the Sky: Millau Viaduct* (documentary). MagellanTV, 2016.

The St. Louis Republican. "The Bridge: The Elephant Declares It a Good Piece of Work— An Immense Crowd of Sight-Seers Yesterday." Monday Morning, June 15, 1874.

Zhia Dao, *History of Transportation in China* (eBook). Kobo, 2019.

These websites were helpful in double-checking bridge specifications:
bridgehunter.com
highestbridges.com
structurae.net

Acknowledgments! The first bridge I remember crossing was the Sherman Minton Bridge (double-decker through arch) connecting the West End of Louisville, Kentucky, to New Albany, Indiana. Love you, Mom and Dad!

Dedication! This book is for Leslie and (our son) Archie, who I hope continues to cross all kinds of bridges, try different things, and make new connections over the span of his entire life.

Join the B.A.T.S.
World-Famous Bridge Brigade!
"WE HANG AROUND BRIDGES!"
BATS
See the world!
Pick out your favorite bridges!
WELCOME BATS
Make new friends!
Or form your own bridge club!
Start making connections today!

SCIENCE
COMICS

GET TO KNOW YOUR UNIVERSE!
SCIENCE COMICS
DINOSAURS
Fossils and Feathers
MK REED
JOE FLOOD

SCIENCE COMICS
CORAL REEFS
Cities of the Ocean
MARIS WICKS

SCIENCE COMICS
VOLCANOES
Fire and Life
JON CHAD

SCIENCE COMICS
BATS
Learning to Fly
FALYNN KOCH

SCIENCE COMICS
FLYING MACHINES
How the Wright Brothers Soared
ALISON WILGUS
MOLLY BROOKS

SCIENCE COMICS
SKYSCRAPERS
The Heights of Engineering
JOHN KERSCHBAUM

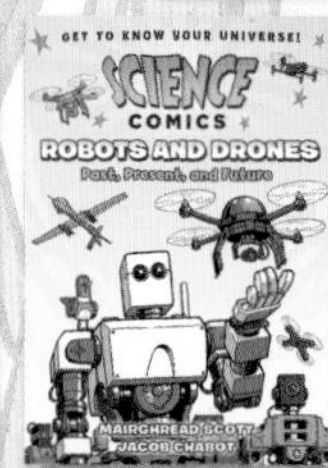
SCIENCE COMICS
ROBOTS AND DRONES
Past, Present, and Future
MAIRGHREAD SCOTT
JACOB CHABOT

SCIENCE COMICS
SHARKS
Nature's Perfect Hunter
JOE FLOOD

SCIENCE COMICS
ROCKETS
Defying Gravity
ANNE DROZD
JERZY DROZD

SCIENCE COMICS
TREES
Kings of the Forest
ANDY HIRSCH

SCIENCE COMICS
SOLAR SYSTEM
Our Place in Space
ROSEMARY MOSCO
JON CHAD

SCIENCE COMICS
THE BRAIN
The Ultimate Thinking Machine
TORY WOOLLCOTT
ALEX GRAUDINS

SCIENCE COMICS
POLAR BEARS
Survival on the Ice
JASON VIOLA
ZACK GIALLONGO

SCIENCE COMICS
WILD WEATHER
Storms, Meteorology, and Climate
MK REED
JONATHAN HILL

SCIENCE COMICS
CROWS
Genius Birds
KYLA VANDERKLUGT

SCIENCE COMICS
CATS
Nature and Nurture
ANDY HIRSCH

SCIENCE COMICS
THE DIGESTIVE SYSTEM
A Tour Through Your Guts
JASON VIOLA
ANDY RISTAINO

SCIENCE COMICS
ROCKS AND MINERALS
Geology From Caverns to the Cosmos
ANDY HIRSCH

SCIENCE COMICS
CARS
Engines That Move You
SCIENCE COMICS
PLAGUES
The Microscopic Battlefield
GET TO KNOW YOUR UNIVERSE!
SCIENCE COMICS
DOGS
From Predator to Protector
ANDY HIRSCH

SCIENCE COMICS
SPIDERS
TAIT HOWARD

SCIENCE COMICS
WHALES
Diving into the Unknown
CASEY ZAKROFF
PAT LEWIS

SCIENCE COMICS
BIRDS OF PREY
Terrible Talons
JOE FLOOD